Modern Approaches in Solid Earth Sciences

Volume 13

More information about this series at http://www.springer.com/series/7377

Andrew Yoram Glikson

The Plutocene: Blueprints for a Post-Anthropocene Greenhouse Earth

The Plutocene is a geological period named after the element Plutonium (Pu) whose radioactive alpha emitting isotope 239Pu of half-life of 24,100 years form a geological marker recording radioactive events.

Springer

Andrew Yoram Glikson
School of Archaeology
and Anthropology
Australian National University
Canberra, Australian Capital Territory,
Australia

Responsible Series Editor: F. Pirajno

Disclaimer: The facts and opinions expressed in this work are those of the author(s) and not necessarily those of the publisher.

Every effort has been made to contact the copyright holders of the figures and tables which have been reproduced from other sources. Anyone who has not been properly credited is requested to contact the publishers, so that due acknowledgment may be made in subsequent editions.

ISSN 1876-1682 ISSN 1876-1690 (electronic)
Modern Approaches in Solid Earth Sciences
ISBN 978-3-319-57236-9 ISBN 978-3-319-57237-6 (eBook)
DOI 10.1007/978-3-319-57237-6

Library of Congress Control Number: 2017941558

Printed on acid-free paper

This Springer imprint is published by Springer Nature
The registered company is Springer International Publishing AG
The registered company address is: Gewerbestrasse 11, 6330 Cham, Switzerland

Mushroom cloud from the Castle Bravo nuclear test in the Bikini Atoll, Marshall Islands. https://en.wikipedia.org/wiki/Castle_Bravo (*Wikipedia, Public Domain*) https://upload.wikimedia.org/wikipedia/commons/5/5d/Castle_Bravo_Blast.jpg

Now I am become Death, the destroyer of worlds.

Bhagavad-gita

---

https://en.wikipedia.org/wiki/J._Robert_Oppenheimer

In Honor of those attempting to defend life,
Including:

Helen Caldicott
Susan Wareham
James Edward Hansen,
Hans Joachim Schellnhuber
William Ernest McKibben
Michael Mann
Albert Arnold Gore
And others

Foreword

The Plutocene is a tour de force, a manuscript that not only has no equal; it hasn't even got any peers, something virtually unknown in today's world of frenetic publication. From end to end a mass of informative illustration take the reader on a manic world-wind tour of humankind, a journey that has no boundaries be they to do with subject matter, time or thought. Poetry, art, history and science are all intermixed with a morass of compressed geological, biological and environmental detail, sometimes in an unlikely combination. Just how this is achieved almost defies imagination but there it's for all to see. The reader is held captive, dragged on in an almost brain-numbing grip as one subject after another comes up and then disappears, to be replaced by something different but equally interesting. This is not a quick read for anybody; it is a book that should be read a bit at a time, with the reader taking long breaks for thought. To cover so much ground the text is economical although jargon words are held in check. Readers with a broad scientific background will have no problems, non-scientists may need google at hand where the book delves into unfamiliar territory or uses unfamiliar terms. That said, Andrew Gilkson's well-honed writing skills come to the fore. Everybody who cares about the future of our planet should read this book and then keep it as a turn-to reference.

Former Chief Scientist,
Australian Marine Institute

Professor J.E.N. (Charlie) Veron

Preamble

On the 27 January, 2017, the Bulletin of the Atomic Scientists moved the arms of its doomsday clock to 2.5 min to midnight, the closest it has been since 1953,[1] with implications for humanity and nature that belong to the unthinkable. In all of the recorded geological history of Earth there is no evidence for an organism perpetrating a mass extinction of species, possibly with the exception of toxic sulphur algae emanating hydrogen sulphide and methane from the oceans (Ward 2007). In a "post-truth"[2] world riddled with "alternative facts", the hapless members of *Homo sapiens*, an intellectually brilliant species, are facing a non-choice of a "*greenhouse summer*" and a "*nuclear winter*". *The physicists have known sin* (Robert Oppenheimer[3]), placing the ultimate weapons in the hands of an imperfect species, for no group can be as wise and in control to indefinitely arrest the effects of lethal carbon emissions and radioactive pollution on the biosphere.

The book "*The Plutocene: Blueprints of a post-Anthropocene Greenhouse Earth*", following on Jonathan Schell's "*The Fate of the Earth*" (Schell 2000) and Clive Hamilton "*A Requiem for a Species*" projects a post-Anthropocene republic of grasses and insects dominated by tropical climate and high radioactivity for at least 20,000 years, the approximate residence time of atmospheric CO_2 and the half-life of Plutonium ^{239}Pu. I tried to make clear separation between the scientific basis for projections of the post-Anthropocene, the natural philosophy arising from these projections, observations of the human condition and poetry—conveying a personal response to humanity's dark night of the soul. As stated by Clive Hamilton" *Sometimes facing up to the truth is just too hard*" (Hamilton 2010).

An exploration of the root causes of the Plutocene in terms of Darwinian evolution theory leads nowhere, nor do theories such as the "*Medea*" (Ward 2015) or "*Cronus*" (Bradshaw and Brook 2006; Sodhi et al. 2009) concepts. Factors arising from the philosophy of science, including the universal nature of intelligence and

[1] https://www.theguardian.com/us-news/2017/jan/26/doomsday-clock-closer-to-midnight-in-wake-of-donald-trump-election?CMP=Share_iOSApp_Other

[2] https://en.oxforddictionaries.com/word-of-the-year/word-of-the-year-2016

[3] http://www.azquotes.com/quote/834918

the vector of purpose in evolution, provide little explanation. While some humans are likely to survive, the winners will be the colonial Arthropods whose tolerance to high temperature and to radiation would ensure their survival. What is left are poetry and Carl Sagan's unfulfilled dictum:

> *We are the local embodiment of a Cosmos grown to self-awareness. We have begun to contemplate our origins: star stuff pondering the stars: organized assemblages of ten billion billion billion atoms considering the evolution of atoms; tracing the long journey by which, here at least, consciousness arose. Our loyalties are to the species and the planet. We speak for the Earth. Our obligation to survive is owed not just to ourselves but also to that Cosmos, ancient and vast, from which we spring*" (Carl Sagan 1980).[4]

Bulletin of the Atomic Scientists 2017.[5]

Canberra, Australia
March, 2017

Andrew Yoram Glikson

[4] https://en.wikipedia.org/wiki/Cosmos_(Carl_Sagan_book)

[5] http://thebulletin.org/current-issue

Acknowledgements

I am grateful to Brenda McAvoy and Darryl Howe for proof reading, Hugh Davies, Ove Guldberg, Gerta Keller, Jon Lomberg, Brian Martin and Charlie Veron for review comments on the manuscript. I thank Charlie Veron for the generous forward. I thank Petra van Steenbergen and Hermine Vloemans for editorial handling. I thank the following people for permission to reproduce figures: Anthony Barnosky, Tom Boden, Gerta Keller, Kay McLaughlin, Robert Nave, Michael Robbins, Elizabeth Sandler and Will Steffen.

I am grateful to Colin Groves and the School of Archaeology and Anthropology, Australian National University, for collaboration in my studies.

Contents

Chapter 1
The Demise of the Holocene Biosphere

1.1 Pandora's Box

"Frontispiece: A painting of Psyche opening the Golden Box" by John William Waterhouse (1903) (https://commons.wikimedia.org/wiki/File:Psyche-Waterhouse.jpg (*Wikipedia commons*))

> *Zeus sent Pandora down to earth and gave her as a present to Prometheus' brother, Epimetheus. Zeus told Epimetheus that he should marry Pandora. Also, Zeus sent Pandora with a little box with a big lock on it (Actually in the earliest versions of this story it is a sealed pottery vase). He said not to ever open the box, and he gave the key to Epimetheus. But Pandora was very curious about what was in the box. She begged Epimetheus to let her open it, but he always said no. Finally one day he fell asleep, and she stole the key (or broke the seal) and opened the box (or vase). Oh! Out of the box flew every kind of trouble that people had never known about before: sicknesses, and worries, and crimes, and hate and*

A.Y. Glikson, *The Plutocene: Blueprints for a Post-Anthropocene Greenhouse Earth*, Modern Approaches in Solid Earth Sciences 13,
DOI 10.1007/978-3-319-57237-6_1

envy and all sorts of bad things. The bad things all began to fly away like little bugs, all over the place. Pandora was very sorry now that she had opened the box! She tried to catch the bad things and put them back in the box but it was too late. They all flew away. But the very last thing to fly out of the box, as Pandora sat there crying, was not as ugly as the others. In fact it was beautiful. It was Hope, which Zeus sent to keep people going when all the nasty things got them down. (https://au.pinterest.com/pin/104708760060078233/)

I had a dream, which was not all a dream.
The bright sun was extinguished and the stars
Did wander darkling in the eternal space,
Rayless, and pathless, and the icy Earth
Swung blind and blackening in the moonless air;
Morn came and went - and came, and brought
No day, and men forgot their passions in the
Dread of their desolation

(From *"Darkness"*, Lord Byron)

Abstract While the global warming train has left the station, with mean temperatures rising toward two degrees Celsius, the nuclear Damocles sword is hanging ever lower over humanity and nature, and the question needs to be asked, what kind of future is *H. sapiens* and numerous species facing? By the initial decades of the twenty-first Century AD, as the atmosphere-ocean-land carbon-oxygen cycle is transformed away from favorable Holocene conditions, which have allowed Neolithic civilization to emerge, critical observations emerge regarding the nature and the consequences of the period defined as the *Anthropocene*, the age of man. Already it is clear the biosphere has entered early to advanced stages of mass extinction of species, progressing at rates one to ten thousand times the pre-industrial rates. Perspectives from paleo-climate science indicate accelerated melting of the large ice sheets, elevated sea level rise rates, an increase in the frequency and intensity of extreme weather events including storms and fires, and an imminent collapse of the Mid Atlantic Ocean Circulation. All along, under false flags, *H. sapiens* continues to develop hair trigger nuclear weapons on the scale costing near two trillion dollars per year, coming out of the mouths of hungry children (http://www.usnews.com/opinion/blogs/robert-schlesinger/2011/09/30/the-origins-of-that-eisenhower-every-gun-that-is-made-quote). Further dissemination of radioactive nuclides on land, oceans and atmosphere heralds the *Plutocene,* a geological era commenced with the Trinity atomic test in 1945 and due to persist for at least 20,000 years, recorded in plutonium-rich clay layers in ocean sediments. Pandora's Box has been opened.

With time possibilities may become probabilities. With time probabilities may become a certainty. By the initial decades of the twenty-first century AD, as *H. sapiens* keeps transforming the carbon-oxygen cycle of the atmosphere and oceans away from the favorable Holocene conditions which allows the Neolithic and civilization to emerge, affecting global warming toward climate tipping points and a major mass extinction of species (Keller 2005), hair-trigger nuclear war heads

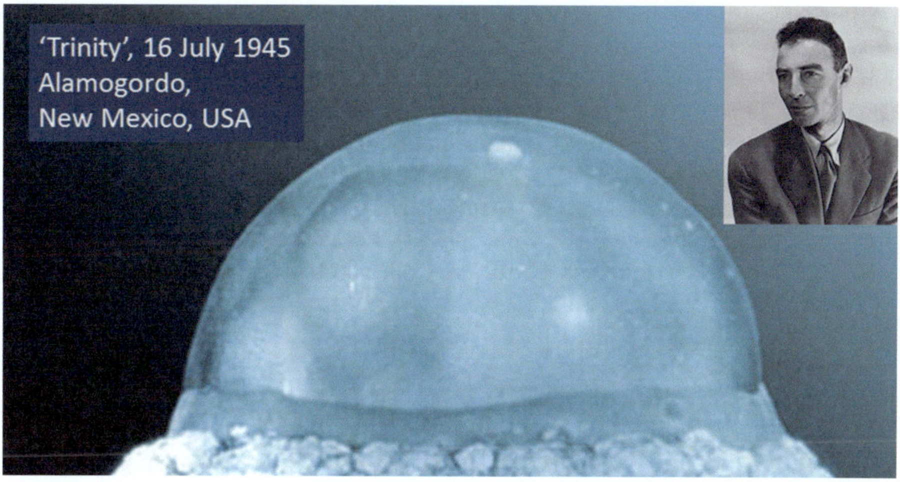

Fig. 1.1 Birth of the *Plutocene*. July 16, 1945. Trinity Site explosion, 0.016 s after the explosion. The viewed hemisphere's highest point in this image is about 200 m high (https://en.wikipedia.org/wiki/Trinity_(nuclear_test)#/media/File:Trinity_Test_Fireball_16ms.jpg) (*Wikipedia, Public Domain*). Inset: Robert Oppenheimer.1904–1967 (https://en.wikipedia.org/wiki/J._Robert_Oppenheimer#/media/File:JROppenheimer-LosAlamos.jpg)

proliferate.[1] It is becoming clear that, instead of channeling its remaining resources in an effort to protect its planetary biosphere, including its own civilization, under false flags *H. sapiens* continues to sink its efforts into murderous enterprises called *war.* This guarantees dissemination of radioactive nuclides on land, oceans and atmosphere, heralding the *Plutocene,* a geological era due to persist for tens of thousands of years, as recorded by the build-up of plutonium-rich clay layers in ocean sediments (Scott et al. 1983).

In the wake of the Late Anthropocoene (Glikson 2013a, 2016) the *Plutocene* was born on July 16, 1945 with the *Trinity* test (Fig. 1.1)[2] and following tests.[3] The *Plutocene* is coined after the element Plutonium as ^{239}Pu decaying to ^{235}U over a half-life of 24,100 years,[4] defining the life-span of the *Plutocene*. During the *Plutocene* the biosphere is dominated by elevated temperatures, analogous to the *Pliocene* (2.6–5.3 Ma ago)[5] or the *Miocene* (~5.3–23 Ma ago)[6] when mean global temperatures were approximately 2–4 degrees Celsius warmer than pre-industrial temperatures, acid oceans and sea levels were approximately 20–40 m higher than

[1] http://www.ploughshares.org/world-nuclear-stockpile-report

[2] https://www.osti.gov/opennet/manhattan-project-history/Events/1945/trinity.htm

[3] https://en.wikipedia.org/wiki/Environmental_radioactivity

[4] https://en.wikipedia.org/wiki/Plutonium

[5] https://www.britannica.com/science/Pliocene-Epoch

[6] https://www.britannica.com/science/Miocene-Epoch

the late Holocene pre-industrial levels (Wynn 2014),[7,8] Atmospheric CO_2 residence time in the order of 10^3–10^4 years (Solomon et al. 2009; Eby et al. 2009) delay the subsequent glacial cycle by perhaps ~50,000 years (Berger and Loutre 2002) (Fig. 1.21). Depending on the extent of nuclear conflicts in the Late Anthropocene, during the *Plutocene* high levels of radioactivity may constrain biological activity to radiation-resistant species.

The abrupt nature of the onset of the *Plutocene* constrains comparisons with conditions of the *Pliocene* environments (USGS 2016) (5.3–2.6 Ma; mean of ~2–3 °C above late Holocene temperatures) and *Miocene* environments (Lewis et al. 2007)[9] (23–5.3 Ma; mean of ~4 °C above late Holocene temperatures). As flora and fauna of the Pliocene and Miocene evolved gradually as compared with the extreme rise of CO_2 in the *Late Anthropocene* at a rate of 3–4 ppm/year,[10] for which there are no precedents as far back as 55 Ma (Zachos et al. 2008), such comparisons only tell part of the story.

In the wake of the *Plutocene,* following a period dominated by grasses and radiation-resistant organisms, mostly arthropods (insects, arachnids, myriapods), a resumption of a glacial period induced by the natural sequestration of CO_2, orbital-driven cooling associated with the Milankovic cycle, and a gradual decline in radioactivity, habitat niches are occupied by survivors, including burrowing mammals. Depending on the intensity of radioactive contamination in different parts of Earth (Rubens et al. 1998), hunter-gatherer humans may survive in northern latitudes and cold high-altitude valleys (World Building 2016).

According to NASA's PLANET (Probing Lensing Anomalies NETwork) project a rough estimate points to the existence of more than 100 billion terrestrial planets across the Milky Way galaxy.[11] A critical parameter in Drake's Equation (SETI 2016), which seeks to estimate the number of planets that host civilizations in the Milky Way galaxy, is L – the longevity of technological societies measured from the time radio telescopes are invented in an attempt to communicate with other planets. Estimates of L vary between 10^2 and 10^7 years, with a value of ~300,000 years proposed as an average (Kompanichenko 2000). Carl Sagan (1980) estimated L on the scale of only a couple of 100 years from the time a civilization discovers nuclear fission, undergoing self-destruction (Sagan 1980). A similar suggestion can be made regarding the onset of combustion of carbon. It is another question whether a species exists in the *Milky Way*, or in any other galaxy, which has triggered a mass extinction of species on the scale initiated by *H. sapiens*.

The history of Earth includes six major mass extinctions which define several periods, including the 850 Ma impact and related extinction and propagation of Acritarch species (Grey et al. 2003; Glikson 2013a, b), end-Ordovician, end-Devonian, Permian-Triassic boundary, end-Jurassic and end-Cretaceous events

---

7 https://www.e-education.psu.edu/earth107/node/901

8 http://www.carbonateresearch.com/node/74

9 https://pubs.usgs.gov/of/2007/1047/ea/of2007-1047ea135.pdf

10 https://www.esrl.noaa.gov/gmd/ccgg/trends/

11 http://hubblesite.org/news_release/news/2012-07

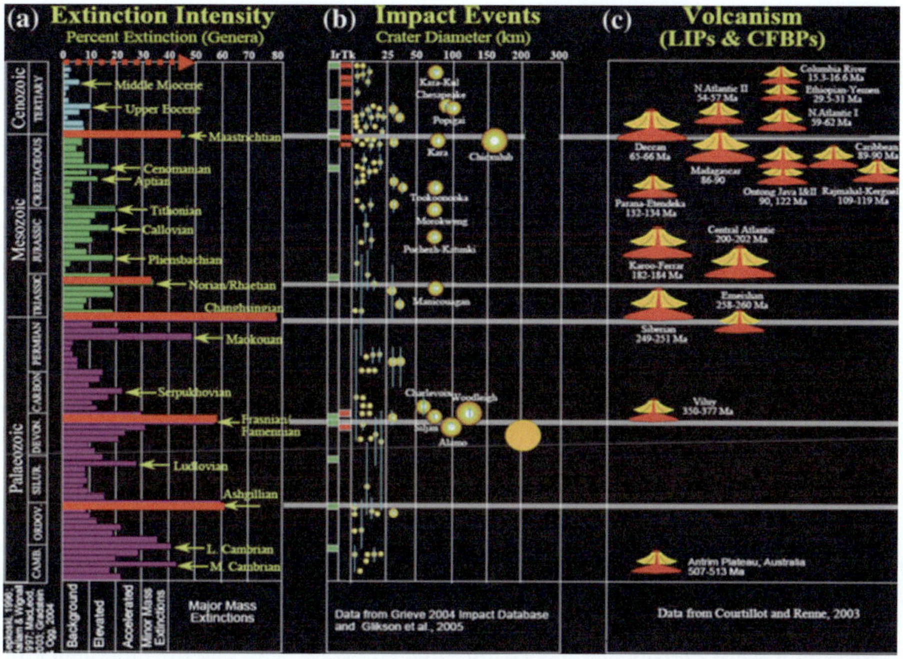

Fig. 1.2 Phanerozoic mass extinctions, asteroid impacts, and large igneous provinces (Keller 2005): (**a**) Extinction intensity; (**b**) Impact events; (**c**) Volcanism. Stratigraphic subdivisions and numerical ages are after Gradstein and Ogg (2004). The extinction record is based on genus-level data by Sepkoski (1998). The number of impact events, size and age of craters follows largely the Earth Impact Database (http://www.passc.net/EarthImpactDatabase/) with modification by the author. The current 7th mass extinction (the 1st mass extinction is defined at 580 Ma at the age of the Acraman impact event with its effect on Acritarchs (Grey et al. 2003) is marked at the top of the figure (Courtesy G. Keller)

(Fig. 1.2) (Glikson 2014; Glikson and Groves 2016; Keller et al. 2012). Each of these events has been triggered by either an extra-terrestrial impact or by massive volcanic eruptions, methane release and related greenhouse events. With the exception of the role of methanogenic bacteria in relation to methane eruptions in the past, the Seventh mass extinction[12] (Fig. 1.2) constitutes a unique event—for the first time in Earth history a rapidly evolving crisis of the biosphere is taking place as a consequence of *biological forcing* by an intelligent carbon-emitting species (Stanley 2016), (Kolbert 2007).

The Milankovic cycles (Roe 2006), including stadial and interstadial glacial-interglacial oscillations during the upper Pleistocene (~10^6–10^4 years-ago), included rapid global and regional temperature changes of about 5 degrees Celsius over a couple of thousand years and in some instances several degrees in a few years (Steffensen et al. 2008), requiring an extreme adaptability of Hominids. Of all life

---

[12] Including the 580 Ma Acraman impact inducing mass extinction of Acritarchs (Grey et al. 2003). http://geology.gsapubs.org/content/31/5/459.abstract

forms only the Genus *Homo* mastered sources of energy— fire, the electromagnetic spectrum and the atom, travelling to other planets, undergoing cultural evolution which has overtaken biological evolution (Midgley 1984). Possessed by a fear of death, craving God-like immortality, *H. sapiens* developed an incongruous faculty of simultaneously creating and destroying, ultimately committing suicide through the destruction of the atmospheric conditions—the lungs of the Earth—which allowed it to develop civilization in the first place. The evolutionary rationales that underlie the transformation of stone-age hunters and gatherers into button-pushing automatons capable of triggering global warming and a nuclear winter remain inexplicable.

About 66 million years-ago two huge asteroids impacting at Chicxulub and Boltysh (Mullen 2004) extinguished some 45% of genera, including the dinosaurs, and vacating habitats for mammals (Alvarez et al. 1980). About 55 Ma, following a sharp rise of atmospheric CO_2 to levels near-1000 parts per million induced by released methane (NASA/Penn State (1991),[13] Earth begun to cool through the sequestration of CO_2 by the rising Himalayan and Alpine ranges, enhancing conditions for warm blooded mammals.[14] About 32 Ma-ago the formation of the large Antarctic ice sheets allowed large warm-blooded mammals to dominate the continents. About 3 Ma-ago, in the mid-Pliocene, when temperatures were about 2 °C higher and sea levels 25 ± 12 metres higher than in the Late Anthropocene, a transition toward Pleistocene glacial-interglacial cycles was saw the emergence of Hominids.[15]

The mastering of fire by *Homo erectus* allowed its long-range migration into cold and arid regions of Earth (Adler 2013). The onset of the Holocene interglacial and stabilization of the climate about 10,000 years-ago, and the cessation of sea level rise about 7600 years-ago, saw the Neolithic take hold, urban civilization, deforestation, ever increasing carbon pollution, acidification of the hydrosphere, radioactive contamination and the demise of species, culminating in the 20th and 21st centuries hallmarks of a new era—the Anthropocoene (Steffen et al. 2007, 2016; Vaughan 2015[16]).

The evidence for a mass extinction of species is summed up by Ceballos et al. (2015) who state: "*Even under our assumptions, which would tend to minimize evidence of an incipient mass extinction, the average rate of vertebrate species loss over the last century is up to 100 times higher than the background rate … these estimates reveal an exceptionally rapid loss of biodiversity over the last few centuries, indicating that a sixth mass extinction is already under way. Averting a dramatic decay of biodiversity and the subsequent loss of ecosystem services is still possible through intensified conservation efforts, but that window of opportunity is rapidly closing*."

[13] https://www.e-education.psu.edu/earth103/node/639

[14] http://geology.gsapubs.org/content/36/12/1003.full

[15] http://www.talkorigins.org/faqs/homs/species.html

[16] https://www.theguardian.com/environment/2015/jun/19/humans-creating-sixth-great-extinction-of-animal-species-say-scientists

Nature includes species whose activities are capable of devastating environments. Toxic viruses, methane and hydrogen sulphide-emitting bacteria, fire ant armies, locust swarms and rabbit populations can lay waste to their surroundings. Host-destroying organisms include species of fungi, worms, arthropods, annelids and vertebrates such as oxpeckers and vampire bats. The mastery of fire enabled the genus *Homo* to magnify its potential to harness and release energy by orders of magnitude, adding its name to the list. From the mid-twentieth century, the splitting of the atom allowed humans to trigger a chain reaction capable of devastating much of the biosphere. A species able to magnify its entropy effect in nature by orders of magnitude would need to be a perfectly wise and controlled species, lest its invention gets out of hand.

1.2 The Lungs of the Earth

Posthumous Life

No one
Was there to hear
The muffled roar of an earthquake,
Nor anyone who froze with fear
Of rising cliffs, eclipsed deep lakes
And sparkling comet-lit horizons
Brighter than one thousand suns
That blinded no one's vision.

No one
Stood there in awe
Of an angry black coned volcano
Nor any pair of eyes that saw
Red streams eject from inferno
Plumes spewing out of Earth
And yellow sulphur clouds
Choking no one's breath.

No one
Was numbed by thunder
As jet black storms gathered
Nor anyone was struck asunder
By lightning, when rocks shuttered
Engulfed by gushing torrents
That drowned the smouldering ashes
Which no one was to lament.

In time
Once again an orange star rose
Above a sleeping archipelago
Sun rays breaking into blue depth ooze
Waves rippling sand's ebb and flow
Receding to submerged twilight worlds
Where budding algal mats
Declare a new life cycle
On the ancient Earth.

Abstract Acting as if the atmosphere constitutes an infinite reservoir into which pollutant can be poured open-ended, the critical role of the oxygen-carbon cycle of the atmosphere has been lost on the majority of humanity, overlooking the evidence that changes in the composition of the oceans and the atmosphere have controlled the history of life on Earth past and present. Acting as the lungs of the biosphere, the physical properties of the troposphere, stratosphere and ionosphere are closely intertwined with photosynthesis and animal biological activity, which in turn exerts an influence on the composition of the atmosphere, by analogy with the relations between marine life and the chemistry of the oceans. Thus past trends of biological evolution and the origin of the great mass extinction can be traced using carbon, oxygen and sulphur isotopes reflecting atmospheric temperatures and CO_2 concentrations and marine acidity levels. The release of more than 600 billion tons of carbon from the Earth crust to the atmosphere since the onset of the industrial age is leading to a fundamental shift in the carbon-oxygen cycle, and thereby the chemistry of the atmosphere and oceans, at a rate exceeding that of the most powerful recorded greenhouse-driven warming events since the 65 million years-old Cretaceous-Tertiary boundary asteroid impact by more than order of magnitude. Remarkably the current carbon emission rate exceeds the carbon release rate of the 56 million years-old Paleocene-Eocene thermal maximum (PETM) and related warming, by a factor of about fifteen, defining Late Anthropocene global warming as an extreme event. The linear to accelerating warming model trends projected by the International Panel for Climate Change (IPCC) are inconsistent with an increase in climate variability driven by amplifying feedbacks, and with an imminent stadial freeze threatened by a collapse of the Atlantic Meridional Ocean Current (AMOC), indicated by a growing cold water region south of the melting Greenland ice sheet.

Forming a thin breathable veneer only slightly more than one thousandth the Earth's diameter (Fig. 1.3), evolving gradually as well as through major perturbations, the atmosphere acts as lungs of the biosphere, facilitating an exchange of carbon and oxygen with plants and animals (Royer et al. 2001, 2004, 2007; Siegenthaler et al. 2005; Berner 2006; Beerling and Royer 2011). The temperature range of the Earth atmosphere (minimum −89.2 °C; mean 15 °C; maximum 56.7 °C)[17] allows the existence of an aqueous medium where metabolic

[17] https://en.wikipedia.org/wiki/Earth

Fig. 1.3 The thin envelope of air that surrounds our planet is a mixture of gases, each with its own physical properties. The mixture is far from evenly divided. Two elements, nitrogen and oxygen, make up 99% of the volume of air. The other 1% is composed of "trace" gases, the most prevalent of which is the inert gaseous element argon. The rest of the trace gases, although present in only minute amounts, are very important to life on earth. Two in particular, carbon dioxide and ozone (Kasting and Donahue 1980), can have a large impact on atmospheric processes (https://www.ucar.edu/learn/1_1_1.htm (University Corporation for Atmospheric Research, NASA))

microbiological processes occur, including extremophile organisms such as chemobacteria around volcanic fumaroles, microbes and algae within and below ice and possibly bacteria in deep crustal fractures. In turn, biological activity continuously modifies the composition of the atmosphere, oceans and soils, for example through production of methane in anoxic environments, release of photosynthetic oxygen from plants and of dimethyl sulphide from marine phytoplankton (PubChem[18]). Long term chemical changes in the air-water system are induced by changes in plate tectonic-driven processes, including volcanic and methane eruptions, weathering, subduction of oceanic plates and continental plates and variations in marine and terrestrial photosynthetic activity (Ruddiman 1997, Raymo and Ruddiman 1992). Long term trends are terminated by abrupt shifts in state of the atmosphere-ocean-ice sheets system (Rahmstorf 2003)[19,20] which may occur at a rate to which many species may be unable to adapt, leading to extinctions. This is the case where the fossil carbon product of ancient biospheres, stored by plants and animals over hundreds of millions of years, are released at the current rates of more than 3 ppm per year (October 2015–398.29 ppm; October 2016–401.57 ppm) (NOAA 2016[21]), threatening the return to the conditions which preceded the emergence of large mammals on land.

---

[18] https://pubchem.ncbi.nlm.nih.gov/compound/dimethyl_sulfide

[19] http://users.clas.ufl.edu/eemartin/GLY6075F08/papers/Broecker06.pdf

[20] http://www.pik-potsdam.de/~stefan/Publications/Journals/rahmstorf_grl_2003.pdf

[21] http://www.esrl.noaa.gov/gmd/ccgg/trends/

1.2.1 Palaeozoic to Cainozoic Climates

An appreciation of the unique nature of climate trends in the *Anthropocene* requires knowledge of the nature of the Earth climate during preceding eras. The appearance of land plants in the late Silurian (~420 Ma), including vascular plants, Cycads and Ginkgo, resulted in an increase in photosynthetic oxygen, which reached a peak of ~31% O_2 in the early Permian (Berner et al. 2007). The build-up of carbon deposits, when combined with photosynthetic oxygen, resulted in flammable land surfaces subject to ignition by lightning and combustion and by volcanic eruptions (Bowman et al. 2009; Vizcarra 2014). The accumulation of carbon residues in sediments over geological periods produced vast reserves of coal and subsequently of petroleum and gas. Early Palaeozoic, early Mesozoic and early Eocene atmospheres saw marked variations in GHG, including stages when CO_2 exceeded 1000 ppm, global temperatures was >5 °C higher than the late Holocene, ensuing in chaotic atmospheric dynamics, intense hydrological cycles and susceptibility to fires (Bowman et al. 2009). The erosion and burial of carbon-bearing sediments, including carbonates, led to a lowering of atmospheric CO_2 (France-Lanord and Derry 1997, Ruddiman 1997) which increasingly exposed the surface to insolation and the Milankovic glacial-interglacial cycles.

The Mesozoic and Cainozoic history of the atmosphere includes marked shifts in its composition between greenhouse gas-rich states (CO_2 = > 1000 ppm) dominated by high-temperature and low pH oceans (Berner 2004, 2006), as in the late Triassic and early Cretaceous, ice ages and cold states, as in the Ordovician-Silurian 450–420 Ma glaciation, late Carboniferous-Permian ~360–260 Ma glaciation, Jurassic and late Cretaceous cold periods and, from 2.6 Ma, the Pleistocene glacial-interglacial (Hansen and Sato 2012). Greenhouse gases are derived from a range of sources, including volcanic activity, bogs, biological activity, methane deposits (Global Carbon Project 2016)[22] and crustal metamorphism. During warm periods low CO_2 solubility in warm oceans led to long-term accumulation of atmospheric CO_2, exceeding 500 ppm and leading to melting of the large ice sheets. Erosion of the Hercynian mountain ranges about 380–280 Ma and of the Tibetan Plateau during the Eocene resulted in CO_2-sequestration, weathering and accumulation of carbonate sediments (Ruddiman 1997; Garzione 2008).

At 64.98 ± 0.05 Ma (Earth Impacts Database)[23] a large asteroid impact event marks the onset of the Cainozoic era (Alvarez et al. 1980; Rennes et al. 2013), followed by a rise from moderate global atmospheric CO_2 concentrations of ~400–~2400 ppm (Beerling et al. 2002). In a wake of a cool middle Paleocene period (Beerling et al. 2002) an abrupt warming episode at 55.9 Ma denoted as the Palaeocene-Eocene thermal maximum (PETM) (Pagani et al. 2006; Zachos et al. 2008; Zeebe et al. 2009) The PETM, lasting approximately ~6.10^3–7.10^3 years, is

[22] http://www.globalcarbonproject.org/methanebudget/

[23] http://www.passc.net/EarthImpactDatabase/NorthAmerica.html

interpreted in terms of the release of approximately ~3000 GtC as methane (Wright and Schaller 2013), elevating CO_2 to ~1800 ppm. Estimates of the rate of carbon release to the atmosphere of about or less than 0.4 ppm CO_2/year are almost an order of magnitude less that the Late Anthropocene rates of above 3 ppm CO_2/year (NOAA 2016[24]). Consequent warming, acidification of the oceans from pH ~8.2 to ~7.5 and oxygen deficiency have markedly affected the marine plankton, including 35–50% of benthic foraminifera (NASA/Pen State).[25] Acidification and warming resulted in extinction of ~35–50% of benthic foraminifera over the course of several thousand years (Zachos et al. 2009; Bowen et al. 2015). Some authors (Cui Ying et al. 2011) estimate the PETM carbon release rate at ~0.3–1.7 GtC/year, considerably less than the near-9.5 GtC in 2012 (Olivier et al. 2014[26]).

According to Abbott et al. (2015) the initial release of carbon during the PETM occurred over at least 4000 years at a rate of 1.1 GtC/year (billion ton carbon per year), whereas anthropogenic release of carbon reached a record high of ~10 GtC/year in 2014 (Zeebe et al. 2016) (Fig. 1.4), an order of magnitude higher than at the outset of the PETM. These authors state: "*We conclude that, given currently available records, the present anthropogenic carbon release rate is unprecedented during the past 66 million years. We suggest that such a 'no-analogue' state represents a fundamental challenge in constraining future climate projections. Also, future ecosystem disruptions are likely to exceed the relatively limited extinctions observed at the PETM*". The consequent rise rate of temperature, analogous to Pleistocene glacial terminations, is exceeded by nearly one order of magnitude during the current late Anthropocene warming rate (Fig. 1.5).

The decline of mean CO_2 and temperatures through the Eocene between ~45 to ~34 Ma was related to a drop in atmospheric CO_2 from >1000 ppm to less than 500 ppm and of mean temperatures by ~5 °C, attributed to CO_2 capture associated with erosion of the rising Himalayan and Alpine mountain chains (Ruddiman 1997). Termination of this cooling was associated with the onset of the Antarctic ice sheet between 34–33.5 Ma, as related to the opening of the Drake Passage between South America and west Antarctica, isolating the Antarctic continent from warmer currents (Zachos et al. 2001). Temperature proxies (oxygen isotopes, Ca/Mg ratios), greenhouse gas proxies (fossil plant pores/stomata), wind proxies (dust), salinity proxies (boron) and plant fossils (pollen, organic remains of algal alkenones, plant wax residues) and other proxies define climate trends of the Oligocene, Miocene and Pliocene (Fig. 1.4)[27,28,29] Low greenhouse rise rates of 10^{-3} ppm CO_2/year and

[24] http://www.esrl.noaa.gov/gmd/ccgg/trends/

[25] https://www.e-education.psu.edu/earth103/node/639; https://en.wikipedia.org/wiki/Paleocene%E2%80%93Eocene_Thermal_Maximum

[26] Olivier JGJ et al (2014) Trends in global CO2 emissions. PBL Netherlands Environmental Assessment Agency, The Hague, p 59

[27] http://users.clas.ufl.edu/eemartin/GLY6075F08/RoyeretalEarthSciRev'01.pdf

[28] http://www.sciencedirect.com/science/article/pii/S0016703706002031

[29] http://uahost.uantwerpen.be/funmorph/raoul/fylsyst/Berner2006.pdf

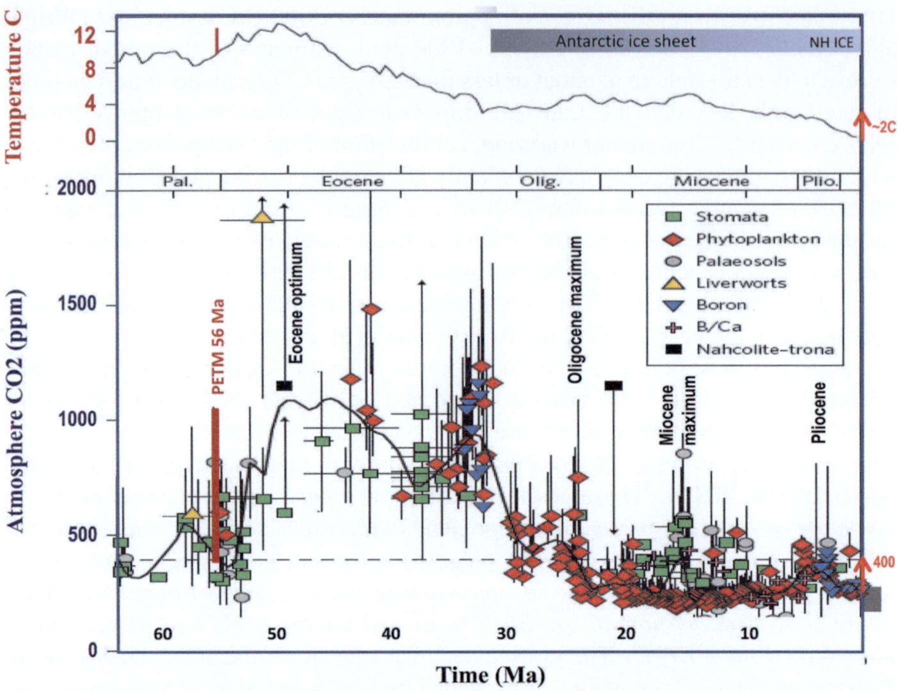

Fig. 1.4 Cenozoic atmospheric CO_2 history by proxy. Deep-sea temperatures (*upper panel*) generally track the estimates of atmospheric CO_2 (*lower panel*) reconstructed from terrestrial and marine proxies following recent revisions. Errors represent reported uncertainties. Symbols with arrows indicate either upper or lower limits. The vertical grey bar on the right axis indicates glacial–interglacial CO_2 range from ice cores. The top blue bar indicates approximate timing of ice-sheet development on Antarctica. Horizontal dashed line indicates the present-day atmospheric CO_2 concentration (390 ppm) (Beerling and Royer 2011) (By permission from Nature Geoscience)

$<10^{-5}$ degrees C/year occur at about ~25 Ma, ~17–14 Ma and ~3.1–2.9 Ma (Zachos et al. 2008). Although short-term climate events may be obscured by the long-term residence time of CO_2, the PETM event is well defined in the geological record,[30] suggesting similar events can be expected to have been detected.

The rise of the Panama Cordillera (Lunt et al. 2008) and thereby the blocking of the Atlantic to Pacific oceanic link between ~13–2.6 Ma was associated with and followed by marked cooling between 2.73 Ma and the first major northern glaciation at 2.15 Ma ago (Rohling et al. 2014). Consequent development of the Pleistocene climate regime, including enhancement of the Pacific Gyre, onset of the ENSO cycle and the North Atlantic Thermohaline Current (NATH), and precipitation over large regions of the Northern Hemisphere led to development of the upper Pleistocene glacial-interglacial cycles. Climates of the last ~800,000 years recorded

[30] http://www.nature.com/nature/journal/v451/n7176/full/nature06588.html

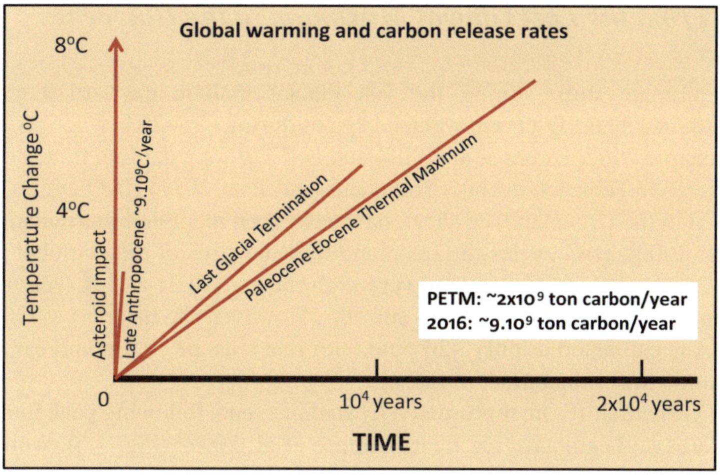

Fig. 1.5 A comparison between the rates of mean global temperature rise during: (1) the last Glacial Termination (after Shakun et al. 2012); (2) the PETM (Paleocene-Eocene Thermal Maximum (after Kump 2011); (3) the late Anthropocene (1750–2016), and (4) an asteroid impact. In the latter instance temperature rise would lag by some weeks or months behind aerosol-induced cooling

in the Greenland and Antarctica ice cores allow detailed information of paleo-CO_2 concentrations and paleo-temperature estimates from Greenland and Antarctica ice cores to 740 kyr, using ice varve count and gas bubbles-trapped CO_2 (Petit et al. 1999, EPICA[31]), allowing detailed resolution of upper Pleistocene climates. The trend was associated with an overall decline of CO_2 from Pliocene levels of approximately ~400 ppm to upper Pleistocene glacial cycles in the range of 180–280 ppm, decline of sea levels by mean of 25 m below Pliocene levels and Pleistocene interglacial levels, and temperature cycles of +/−5 °C related to orbital forcing of the Milankovic glacial-interglacial cycles (Hansen et al. 2007, Hansen and Sato 2012). Glacial phases were associated with development of the Laurentian, Greenland and Fennoscandian ice sheets and the replacement of rainforests with open savanna (de Menocal 2004). This included an increase in climate variability as marked by the diversifications of antelopes at ~2.8 Ma, 1.8–1.7 Ma and 0.8–0.7 Ma, and a variability selection of Hominids from the Olduvan stone tools age from ~2.7 Ma to the Acheulean stone tools age from ~1.7 Ma (Klein and Edgar 2002). Glacial phases intervened by interglacial phases were triggered by solar pulsations maxima of ~40–60 Watt/m^2 in high northern latitudes (~65 N) amplified by the ice/water albedo flip process (Hansen et al. 2008, Hudson 2011) and further release of CO_2 from warming water, lagging by approximately ~700 years behind temperature.

[31] http://cdiac.ornl.gov/trends/co2/ice_core_co2.html

1.2.2 From the Last Glacial Maximum to the Holocene

Extreme climate shifts during the last glacial include the ~1470 years-long Dansgaard-Oeschger (D-O) intra-glacial cycles during ~75–20 kyr-ago (Ganopolski and Rahmstorf 2002; Braun et al. 2008) and 6–7 kyr-long cooling trends culminating in so-called Heinrich events (Yokoyama and Esat 2011). D-O cycles, initially recognised in the Greenland ice sheet, have also been identified in tropical latitudes (Broecker 2000). D-O cycles in Greenland indicate rises of mean global temperatures by about ~3 °C at moderate to very high rates of ~0.01–0.2 °C/year related to insolation peaks amplified by ocean currents. However, the increase in CO_2 during these rises is estimated as only ~20 ppm with rise rates of ~0.2 ppm/year, an order of magnitude lower than modern rates.

Major climate shifts are represented by stadial events following peak inter-glacial temperatures and extensive ice melt (Cortese et al. 2007) (Fig. 1.6), with a prime example represented by the 12.9–11.7 kyr *Younger Dryas* phase (Fig. 1.7) in Greenland displaying abrupt freezing over tipping points as short as 1–3 years (Steffensen et al. 2008). A younger stadial at ~8.2 kyr is represented by a freeze in the North Atlantic region in the wake of extensive melting of the Laurentian ice sheet, flowing to large North American lakes (Lake Agassiz). The 8.2 kyr freeze gave way to the *Holocene Optimum* ~8.0 kyr-ago followed by a decline of CO_2 and methane to about ~6000 BC and of methane from ~4000 BC. According to Ruddiman (2003) and Kutzbach et al. (2010) long-term cooling was interrupted by release of greenhouse gases due to Neolithic land clearing and burning. Other authors interpret this rise as a natural perturbation (Broecker 2006; Broecker and Stocker 2006). Ganopolski et al. (2016) suggest a decline to a new ice age was missed even prior to the onset of the industrial age and cumulative CO_2 emissions of 1000–1500 GtC may delay the onset of the next glacial inception by at least 100,000 years. The Holocene record includes a relatively mild regional warming stage (Medieval Warm Period ~900–1400 AD) and a cool phase (Little Ice Age ~ 1650–1600 AD). These events corresponded respectively to high solar radiation levels of +0.5 watt/m^2 and low solar activity of −0.5 watt/m^2, the latter associated with a near cessation of sun spot activity (Solanki 2002).

1.2.3 An Orwellian Climate

From 1870 to 2014, cumulative carbon emissions totalled about 545 GtC (Figs. 1.8 and 1.9). The global consequences of these emissions in terms of their intensity and partitioning of CO_2 between the atmosphere, land and ocean (Fig. 1.10) and in terms of fires (Fig. 1.11), leaf growth (Fig. 1.12) are portrayed below. The consequences of temperature rise (Figs. 1.13, 1.20 and 1.21) are portrayed in terms of melting of the Arctic sea ice and Greenland ice sheet and their consequences (National Academies Press 2017) (Figs. 1.14 and 1.15), the Antarctic ice sheet (Fig. 1.16) and in terms of sea level rise (Figs. 1.17 and 1.18).

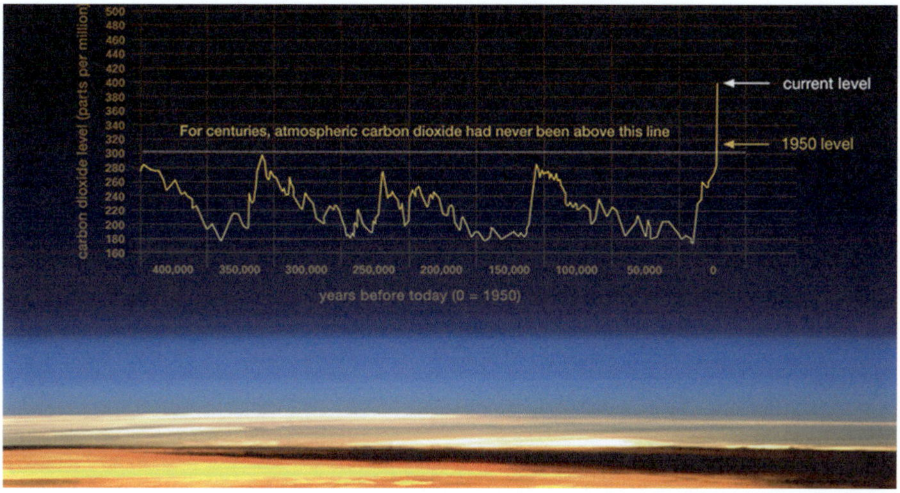

Fig. 1.6 The carbon cycle resent and past: Evolution of CO_2 over the last 400,000 years (NOAA https://climate.nasa.gov/climate_resources/24/)

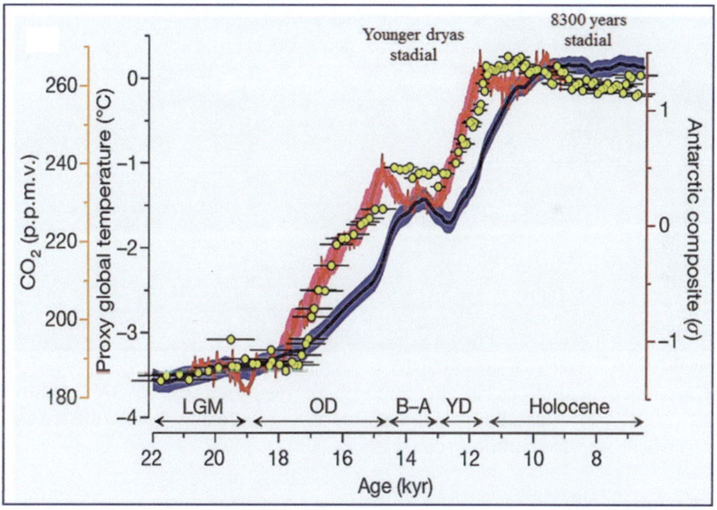

Fig. 1.7 The global proxy temperature stack (*blue*) as deviations from the early Holocene (11.5–6.5 kyr ago) mean, an Antarctic ice-core composite temperature record (*red*), and atmospheric CO_2 concentration. The Holocene, Younger Dryas (*YD*), Bølling–Alerød (*B–A*), Oldest Dryas (*OD*) and Last Glacial Maximum (*LGM*) intervals are indicated. p.p.m.v., parts per million by volume (Shakun et al. 2012) (Nature, by permission)

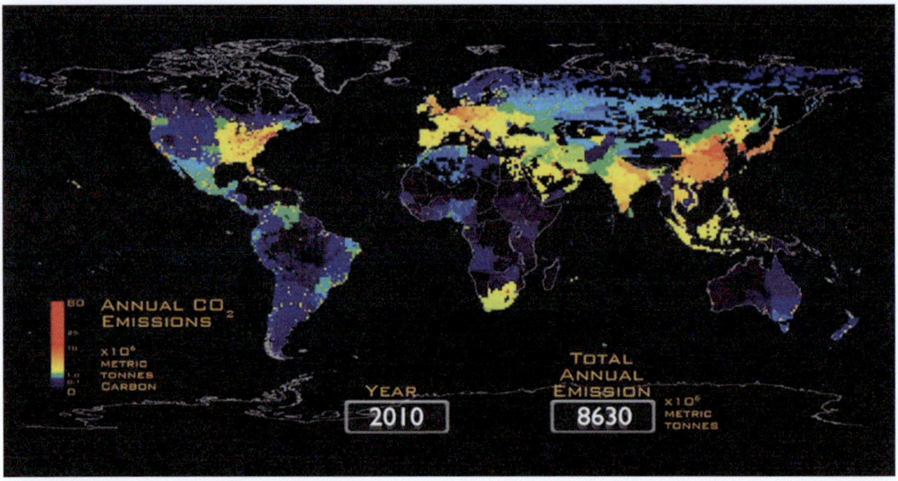

Fig. 1.8 Annual carbon emissions for 2010 (By permission, Tom Boden, Carbon Dioxide Information Analysis Center, Oak Ridge National Laboratory) (http://www.globalcarbonproject.org/carbonbudget/16/media.htm)

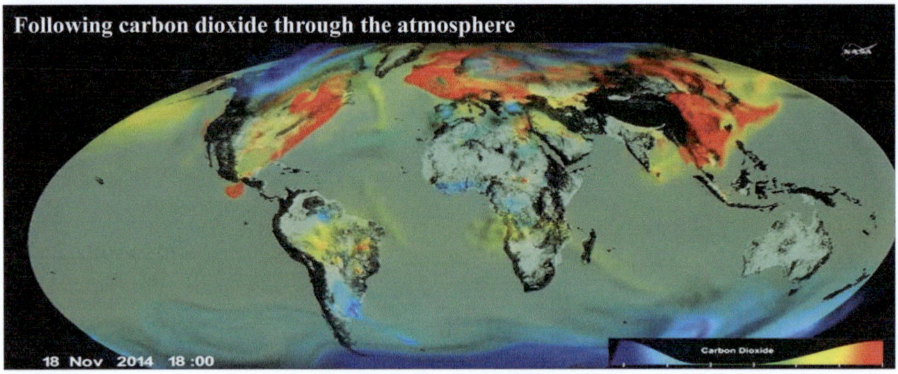

Fig. 1.9 Carbon concentrations in the atmosphere measured by NASA's Orbiting Carbon Observatory (OCO-2) satellite (https://www.nasa.gov/feature/goddard/2016/eye-popping-view-of-co2-critical-step-for-carbon-cycle-science)

Emissions were partitioned among the atmosphere (approx. 230 GtC or 42%), ocean (approx. 155 GtC or 28%) and the land (approx. 160 GtC or 29%) (CO_2-Earth 2015).[32] Currently rising at approximately 9 GtC/year, this constitutes a fundamental shift in the state of the terrestrial atmosphere (Figs. 1.20 and 1.21). Emissions in 2014 were derived from coal (42%), oil (33%), gas (19%), cement (6%) and gas flaring (1%) and are partitioned between the atmosphere (42%), the ocean (28%)

[32] https://www.co2.earth/global-co2-emissions

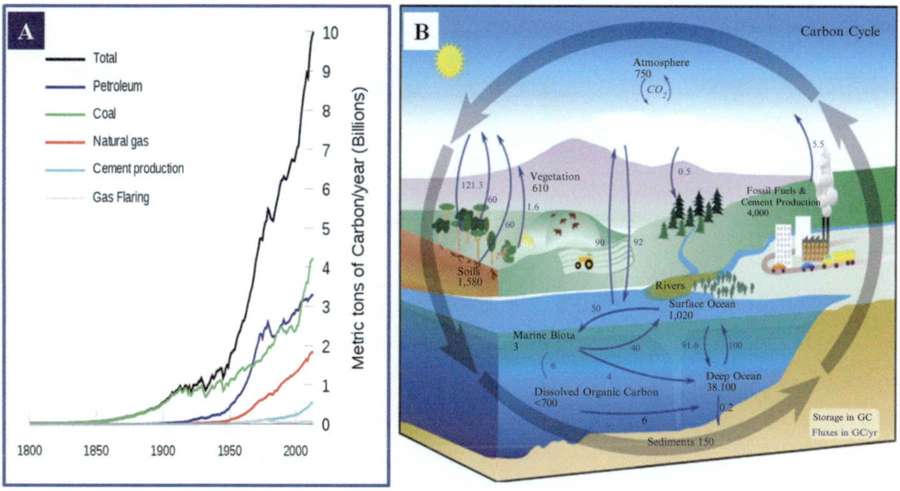

Fig. 1.10 (**a**) Global annual fossil fuel carbon dioxide emissions through year 2007, in billion metric tons of carbon, specified by fuel type and source of emission (https://commons.wikimedia.org/wiki/File:Global_Carbon_Emissions.svg); (**b**) Storage and annual exchange of carbon between the atmosphere, hydrosphere and geosphere in gigaton of Carbon (GtC). (https://commons.wikimedia.org/wiki/File:Carbon_cycle-cute_diagram.svg). 2016 global emissions ~10 GtC/year (https://www.carbonbrief.org/what-global-co2-emissions-2016-mean-climate-change)

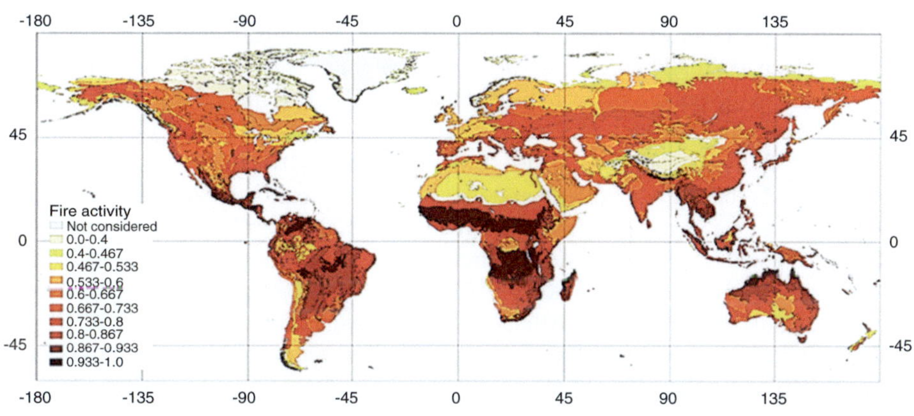

Fig. 1.11 Ecologically-based global fire map derived from fire activity data, Net Primary Productivity (NPP) data, and Normalized Difference Vegetation Index (NDVI) data. NASA (https://earthdata.nasa.gov/user-resources/sensing-our-planet/strange-bedfellows-2014)

and the land (29%).[33] Changes in land use are responsible for about 9% of all global CO_2 emissions. In 2013 emission growth originated from China (58%), USA (20%) and India (17%).[34]

[33] http://cdiac.ornl.gov/GCP/

[34] http://www.globalcarbonproject.org/carbonbudget/16/media.htm

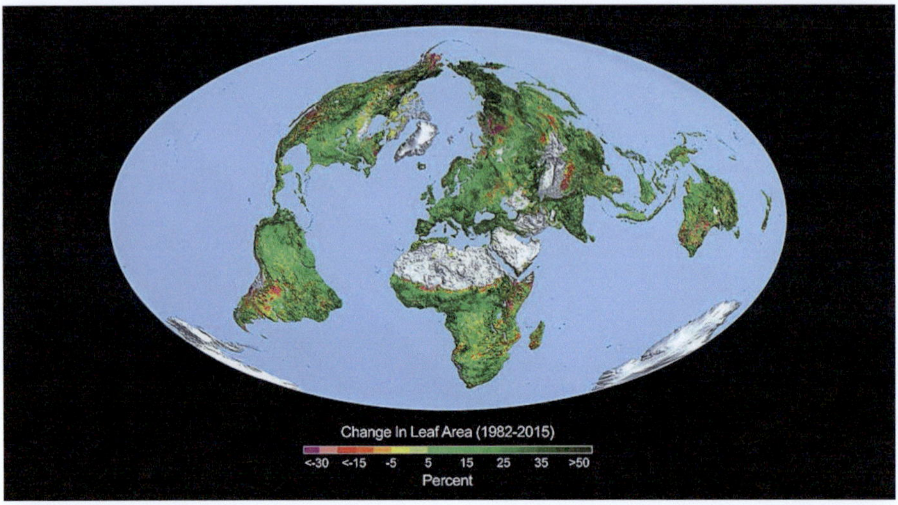

Fig. 1.12 This image shows the change in leaf area across the globe from 1982–2015. (NASA) (Credits: Boston University/R. Myneni https://www.nasa.gov/feature/goddard/2016/carbon-dioxide-fertilization-greening-earth)

The rise in carbon gases has increased the radiation of the atmosphere by about +3.2 Watt/m^2, which, for a climate sensitivity value of 3 °C for doubling of CO_2 translates to ~2.4 °C (Hansen et al. 2011). A large part of the radiation is masked by the clouding effects of sulphur aerosols. The mean global temperature rise of +1.3 °C since 1880 AD is twice as strong on the continents as in the oceans and is accentuated in the polar regions to higher than +5 °C due to the ice/water albedo flip (Hansen et al. 2007, 2011).

Post-1750 Atmospheric CO_2 rise from ~280 to >401 ppm, with a mean of 0.45 ppm/year, has accelerated to a rate near 3.3 ppm/year between October 2015–2016,[35] exceeding any in the geological record, with the possible exception of CO_2 release triggered by the K-T asteroid impact (Beerling et al. 2002, Glikson 2016). Mean temperature rise rates of ~0.009 °C/Year since 1880 exceed all recorded previous rates, excepting regional Atlantic Ocean Dansgaard-Oeschger rates of 0.01–0.2 °C/year (Ganopolski and Rahmstorf 2002). The steep rise in carbon emissions since WWII and a sharp rise of mean global temperatures from about 1975–76 signify an abrupt shift in state of the atmosphere-ocean climate system (Fig. 1.6, 1.20 and 1.21). Mild cooling during 1940–1975 was related to rising levels of attenuating sulphur aerosols and a low in the 11-years sun spot cycle. Likewise lowering of the rate of temperature rise from about 1999–2000 can be attributed to attenuation by the increase of sulphur aerosol emission Smith et al. (2011) and low sunspot activity (NASA 2016[36]). Peak temperatures of +0.62 °C and +0.63 °C relative to 1951–1980 were reached in 2005 and 2010. A development of a cool water region south of

[35] http://www.esrl.noaa.gov/gmd/ccgg/trends/

[36] https://solarscience.msfc.nasa.gov/SunspotCycle.shtml

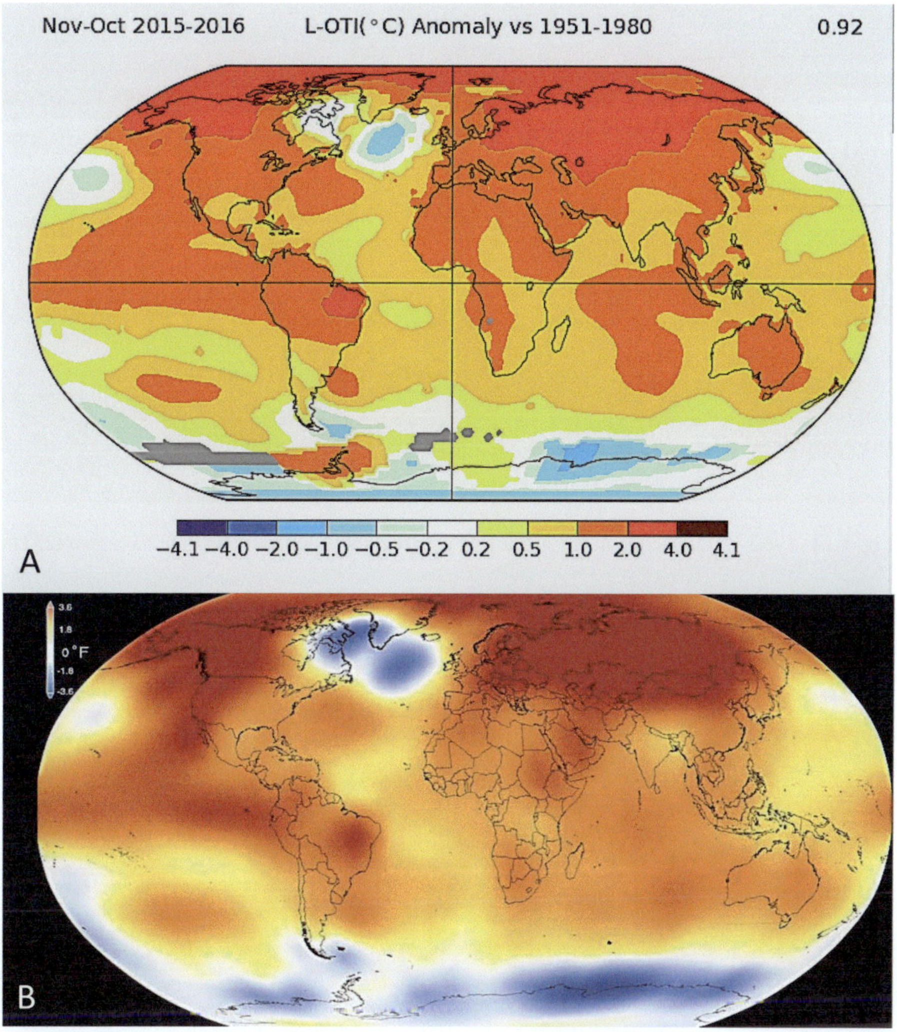

Fig. 1.13 Global warming: (**a**) 2015 was the warmest year since modern record-keeping began in 1880, continuing a long-term warming trend—15 of the 16 warmest years on record have occurred since 2001 (NASA http://data.giss.nasa.gov/gistemp/maps/); (**b**) November 2015 to October 2016 mean temperatures (NASA https://www.nasa.gov/press-release/nasa-noaa-analyses-reveal-record-shattering- global-warm-temperatures-in-2015)

Greenland (Fig. 1.13) signifies a future collapse of the Atlantic Meridional Ocean Current (Rahmstorf et al. 2015). At an unspecified stage of global warming, possibly before temperatures reach 2 °C above pre-industrial levels, a collapse of the Atlantic Meridional Overturning Circulation (AMOC)[37] would result in extensive freeze conditions around the Atlantic, perhaps analogous to stadial events such as

[37] http://www.nature.com/nclimate/journal/v5/n5/full/nclimate2554.html

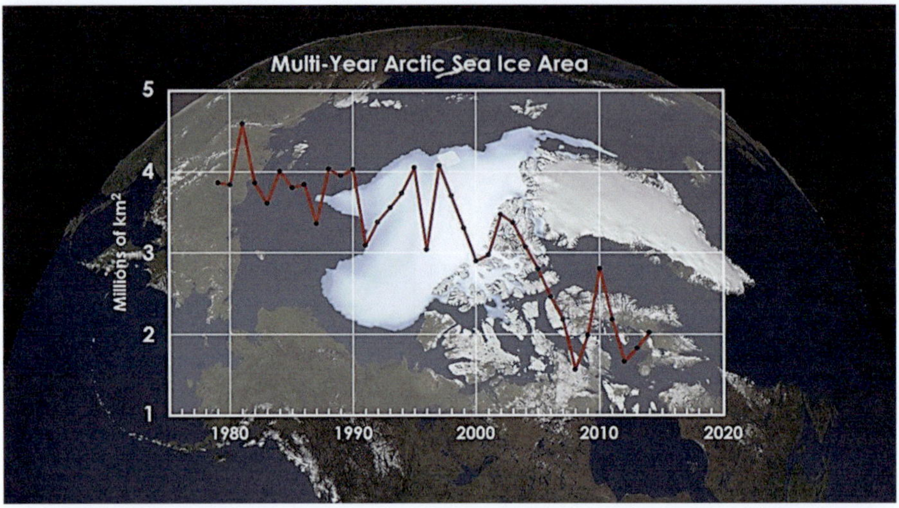

Fig. 1.14 1980–2015 Decline in the Arctic Sea ice area. (NASA https://svs.gsfc.nasa.gov/4251)

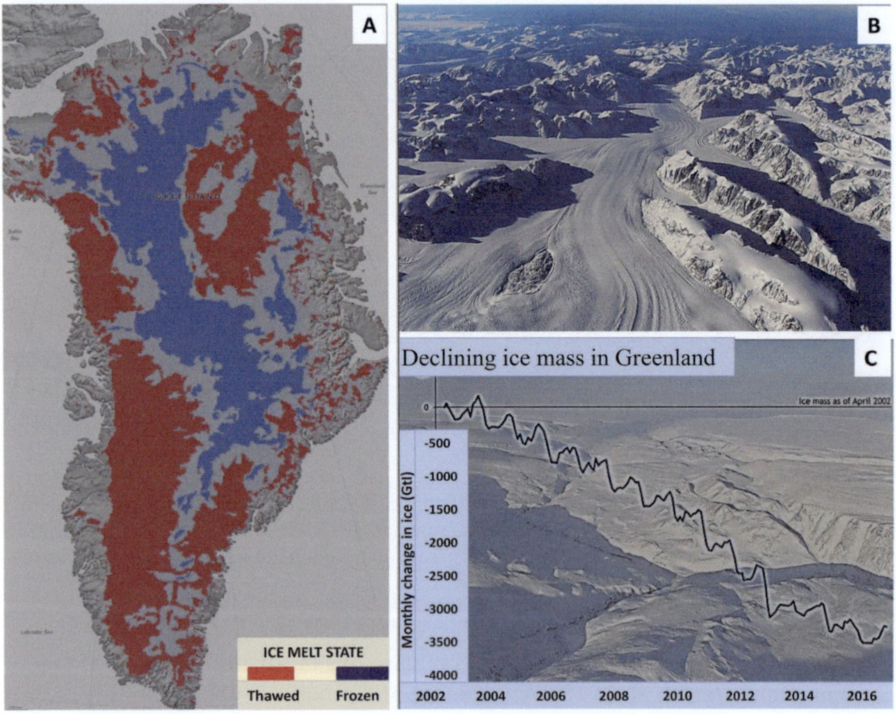

Fig. 1.15 (**a**) State of Greenland ice in 2016. Thawed (*red*) and frozen (*blue*) ice (https://www.nasa.gov/sites/default/files/thumbnails/image/greenlandmelt_basalmelt_1041.jpg); (**b**) Greenland glacier (https://www.nasa.gov/sites/default/files/thumbnails/image/heimdal_0.jpg); (**c**) Declining ice mass between 2002–2016 (https://www.climate.gov/news-features/featured-images/greenland-ice-mass-loss-continued-2016)

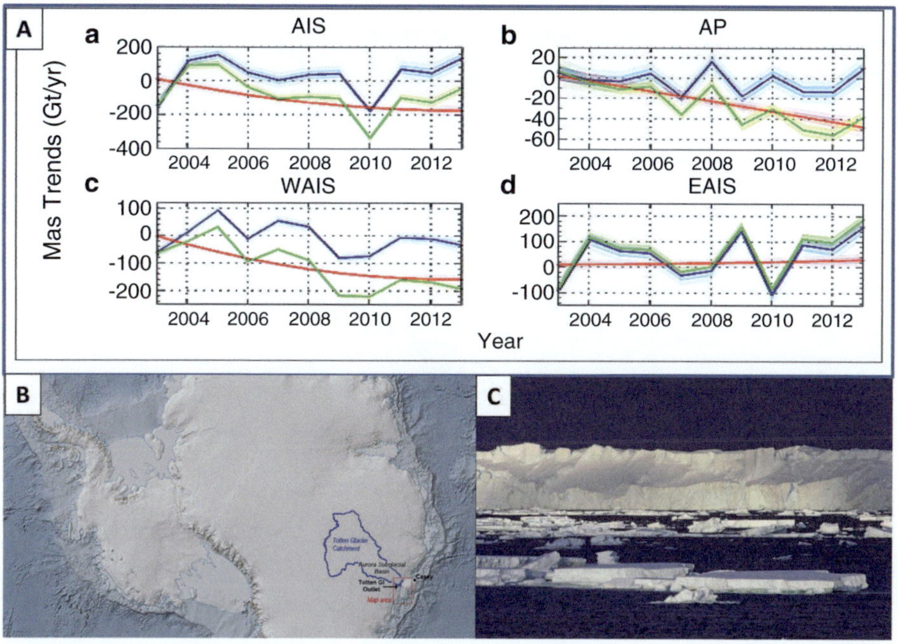

Fig. 1.16 (**a**) Antarctic ice change trends (*AIS* Antarctic ice sheet, *AP* Antarctic Peninsula, *WAIS* West Antarctic ice Sheet, *EAIS* East Antarctica Ice Sheet) (http://onlinelibrary.wiley.com/doi/10.1002/2015JF003550/full; http://publications.agu.org/author-resource-center/usage-permissions/#repository); (**b**) Antar4ctica sea ice, showing the area of the Totten Glacier (https://www.nasa.gov/sites/default/files/thumbnails/image/earth20150316c. jpg); (**c**) The Totten Glacier (http://www.abc.net.au/news/2016-12-17/search-for-source-of-warm-water-melting-totten-glacier/8129202)

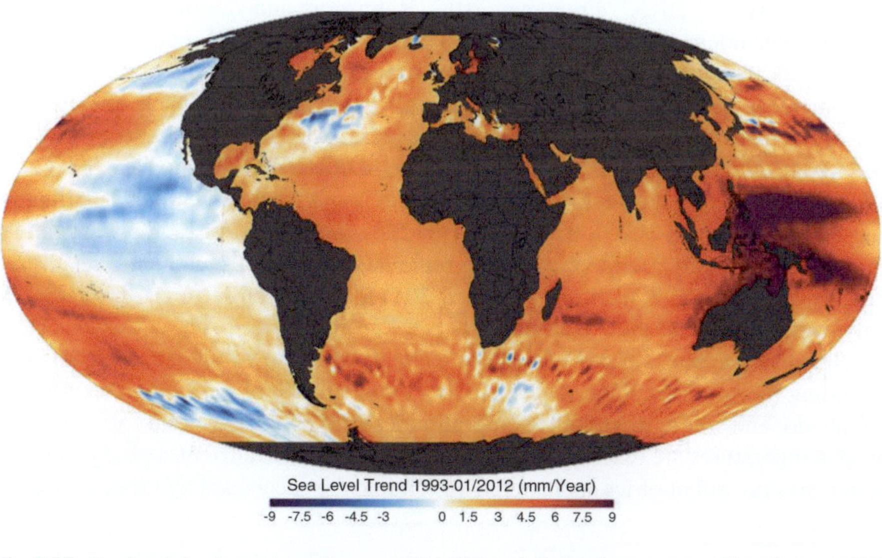

Fig. 1.17 Sea level rise in mm/year January 2013 (NOAA) (https://commons.wikimedia.org/wiki/File:NOAA_sea_level_trend_1993_2010.png)

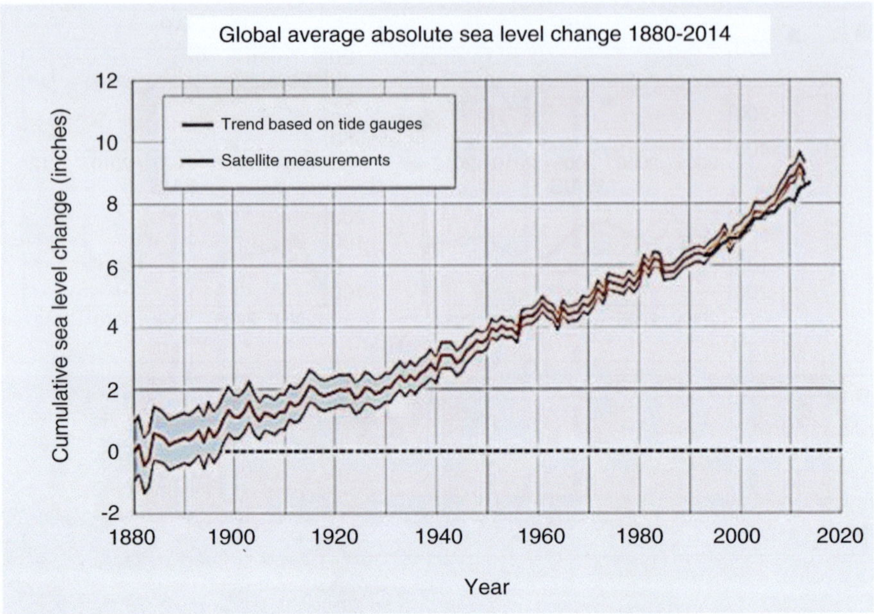

Fig. 1.18 Sea level rise 1880–2013 (NOAA) (https://commons.wikimedia.org/wiki/File:Trends_in_global_average_absolute_sea_level,_1880-2013.png)

triggered 8500 years-ago by Laurentian ice melt (Wagner et al. 2002) and the *Younger dryas* 12,700–11,900 years ago (Carlson 2010).

Rising temperatures drive an increase in evaporation from the oceans, enhancing the hydrological cycle as represented by intensification of cyclones, major precipitation events, floods, pole-ward shift in the storm tracks and related disruption of vegetation (IPCC 2013,[38] Freedman 2012, Rahmstorf and Coumou 2011). Rising ocean water temperatures decreases CO_2 solubility, lowers pH and decreases biological calcification and thereby CO_2 sequestration. Desiccated and burning vegetation increase release of CO_2 to the atmosphere. A major feedback effect is the release of methane since 1750 affecting an increase in the global energy level of ~0.5 Watt/m^2 (Hansen et al. 2012), from Arctic permafrost (estimated as ~900 GtC), high-latitude peat lands (estimated as ~400 GtC) and tropical peat lands estimated as ~100 GtC (Scope 2006). Between 1979–2007 the loss of Arctic ice was equivalent to +0.1 Watt.m^{-2} while a complete removal of Arctic sea ice would result in a forcing of about +0.7 Watt.m^{-2} (Hudson 2011), a value close to Earth's radiative imbalance in 2011 of ~0.6 Watt/m^2.

Negative feedbacks occur when cold ice melt water flow into ocean regions surrounding the continental ice sheets, leading to the growth of sea ice and an increase of albedo, and where increasing evaporation leads to enhanced cloud formation. As peak temperatures are reached they may be followed by a transient stadial phase due to the cooling effect of ice melt water, which may be succeeded by renewed warm-

[38] http://climatechange2013.org/

ing consequent on the rising concentrations of CO_2 in the atmosphere (Solomon et al. 2009). The consequences of global warming include shifts in storm tracks and collision between cold and warm air masses affecting weather patterns, in particular in the Northern Hemisphere. The poleward shift of arid zones results in drying up of temperate zones, for example in northwest China, North Africa, southern Europe, south and southwest Australia and southern Africa, where forests are affected by heat waves and firestorms. The expansion of tropical climate zones leads to greening of marginal desert zones and in the northern hemisphere migration of the tundra toward the Arctic.[39] Global temperature, ice melt and sea levels lag behind atmospheric radiative forcing, in part masked by sulphur aerosols (Hansen and Sato 2012). Melting of the large ice sheets of West Antarctica and Greenland has been doubling every 5–10 years (Velicogna 2014; Velicogna and Wahr 2005; Espanol et al. 2016), driving sea level rise at a rate faster than that induced by thermal expansion and mountain glaciers. Greenland mass loss has risen from 137 GtI/yr. in 2002–2003 to 286 GtI/yr. in 2007–2009. Mass ice loss in Antarctica increased from 104 GtI/yr. in 2002–2006 to 246 GtI/yr. in 2006–2009 (Velicogna 2009, Rignot et al. 2011).

Sea levels constitutes the definitive manifestation of climate change processes, including thermal expansion or contraction of water and melting or freezing of ice sheets and mountain glaciers, and may lag behind rising radiative forcing levels. Studies of temperature-ice volume-sea level relations for five periods during the last 40 Million years manifest a dominant role of CO_2 in controlling ice volumes and sea levels (Foster and Rohling 2013). Where greenhouse gases reach a level of 400–450 ppm they are likely to lead to sea levels rise of many meters above those in the early twenty-first century (Hansen 2007, 2012).

However, the effects of ice melt on the oceans are complex (Rhiannon 2016). According to Hansen et al. (2016) "*the effect of growing ice melt from Antarctica and Greenland tends to stabilize the ocean column, inducing amplifying feedbacks that increase subsurface ocean warming and ice shelf melting … thus increasing Earth's energy imbalance and heat flux into most of the global ocean's surface. Southern Ocean surface cooling, while lower latitudes are warming, increases precipitation on the Southern Ocean, increasing ocean stratification, slowing deep-water formation, and increasing ice sheet mass loss. These feedbacks make ice sheets in contact with the ocean vulnerable to accelerating disintegration … Doubling times of 10, 20 or 40 years yield multi-meter sea level rise in about 50, 100 or 200 years … the feedbacks, including subsurface ocean warming, help explain paleoclimate data and point to a dominant Southern Ocean role in controlling atmospheric* CO_2*, which in turn exercised tight control on global temperature and sea level. The millennial (500–2000-year) timescale of deep-ocean ventilation affects the timescale for natural* CO_2 *change and thus the timescale for paleo-global climate, ice sheet, and sea … These climate feedbacks aid interpretation of events late in the prior interglacial (early Holocene) when sea level rose to 6–9 meters with evidence of extreme storms while Earth was less than 1 °C warmer than today. Ice melt cooling of the North Atlantic and Southern oceans increases atmospheric temperature gra-*

[39] https://www.ipcc.ch/publications_and_data/ar4/wg2/en/ch4s4-4-10.html

dients, eddy kinetic energy and baroclinicity (misalignment of pressure gradients), thus driving more powerful storms. The modeling, paleoclimate evidence, and ongoing observations together imply that 2 °C global warming above the preindustrial level could be dangerous. Continued high fossil fuel emissions this century are predicted to yield (1) cooling of the Southern Ocean, especially in the Western Hemisphere; (2) slowing of the Southern Ocean overturning circulation, warming of the ice shelves, and growing ice sheet mass loss; (3) slowdown and eventual shutdown of the Atlantic overturning circulation with cooling of the North Atlantic region; (4) increasingly powerful storms; and (5) nonlinearly growing sea level rise, reaching several meters over a timescale of 50–150 years. These predictions, especially the cooling in the Southern Ocean and North Atlantic with markedly reduced warming or even cooling in Europe, differ fundamentally from existing climate change assessments." CO_2-equivalent of above >460 ppm, which includes the effects of methane and nitrous oxide, commit the atmosphere to a warming trend approaching the upper stability level of 500+/−50 ppm CO_2 of the Antarctic ice sheet (Zachos et al. 2001).

At 0.003–0.016 °C/year modern temperature rise rates are high by an order of magnitude as compared to the glacial terminations (Eemian ~0.0004 °C/year), whereas the modern sea level rise rate of ~3.5 mm/year is similar to or lower than during glacial terminations of ~3–10 mm/year, suggesting a lag effect (Glikson 2016). Equilibrium relations between a temperature rise of 2–3 °C above pre-industrial levels and sea level rise implies a rise toward Pliocene-like sea levels, estimated at 4–40 m above the present (Raymo 2009) modelled at ~25 m (Raymo et al. 2011; Chandler et al. 1994) and a permanent El-Nino state (Haywood et al. 2007). Due to ice sheet hysteresis the precise time table for future sea level rise remains unknown.

The mostly linear to accelerating trends portrayed by climate models published by the International Panel of Climate Change (IPCC 2007a, b) are contrasted with the abrupt shifts in state and tipping points during the glacial-interglacial cycles and even Holocene history (Steffensen et al. 2008[40]). Examples are the stadial freeze and recovery events (Figs. 1.6 and 1.7) amplified by feedback effects induced by the ice-water albedo flip effects (IPCC 2016[41], Lamont Doherty Observatory 2016[42]). Tipping points in the climate (Lenton et al. 2009) occur when temperature changes due to solar radiation or greenhouse gas levels reach points where positive or negative feedback render changes irreversible. Such climate shifts comprise melting and collapse of Greenland and west Antarctic ice sheets, permafrost melting, boreal forest and tundra loss, monsoon shifts, Amazon forest dieback (Lenton et al. 2009), changes in the ENSO circulation, deep water formation and extreme weather events (Hansen et al. 2012, Rahmstorf and Coumou).

To date there has been limited appreciation of the full consequences of abrupt shifts in the CO_2 and Oxygen composition on the biosphere. An analogy can be drawn with the consequences of changes to the CO_2 and O_2 concentrations in the

---

[40] http://science.sciencemag.org/content/321/5889/680

[41] https://www.ipcc.ch/ipccreports/tar/wg1/074.htm

[42] https://www.sciencedaily.com/releases/2016/06/160630145000.htm

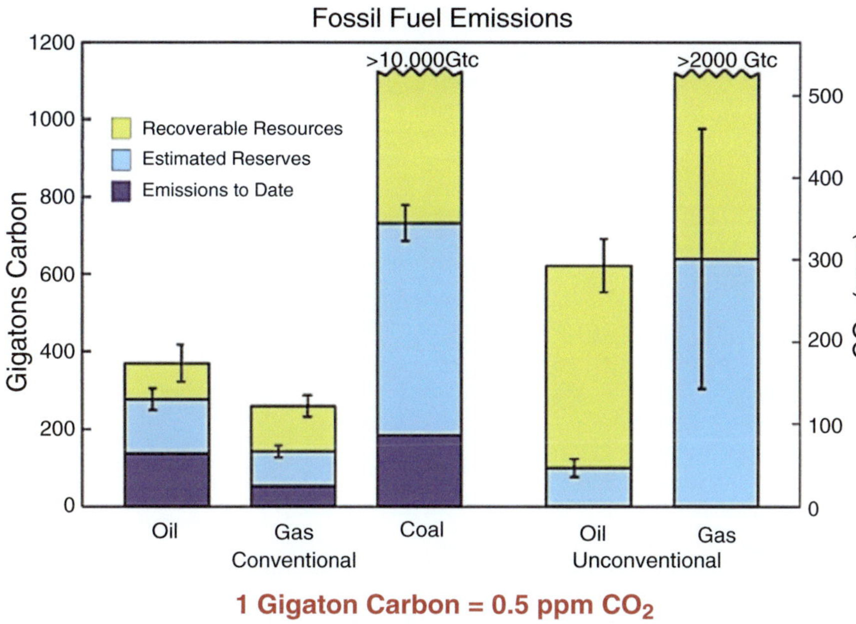

Fig. 1.19 Estimates of fossil fuel resources and equivalent atmospheric CO_2 levels, including (1) emissions to date; (2) estimated reserves, and (3) recoverable resources (1 ppm CO_2 ~ 2.12 GtC) (Hansen et al. 2012) (http://journals.plos.org/plosone/article?id=10.1371/journal.pone.0081648; http://journals.plos.org/plosone/s/licenses-and-copyright)

lungs of animals and humans, leading to hypercapnia (Kregenow and Swenson 2002).[43] The residence time of CO_2 in the atmosphere of about 10^3–10^4 years (Solomon et al. 2009, Eby et al. 2009) implies a long-term shift in the onset of the next ice age, possibly by tens of thousands of years. Multi-millennial projections suggest the long-term climate response depends on the cumulative total amount of emissions (Eby et al. 2009). For emissions similar to the known and predicted fossil fuel reserves (Fig. 1.19) the time to absorb 50% of the CO_2 would exceed 2000 years and the lifetime of the surface air temperature anomaly might be 60% longer than the lifetime of anthropogenic CO_2. A rise of atmospheric CO_2 at the current rate of >3 ppm/year above 400 ppm and toward levels of 600 ppm would result in global temperatures 4 °–6 °C higher than pre-industrial temperatures (Fig. 1.20), further extensive melting of the large ice sheets and sea level rise on the scale of many of meters.[44] Two-thirds of the maximum temperature anomaly will persist for longer than 10,000 years, suggesting the consequences of anthropogenic CO_2 emissions lead to tropical Earth conditions lasting for millennia.[45]

[43] http://erj.ersjournals.com/content/20/1/6

[44] https://pubs.giss.nasa.gov/abs/ha01210n.html

[45] http://web.maths.unsw.edu.au/~katrinmeissner/pdfs/eby_jc_2009.pdf

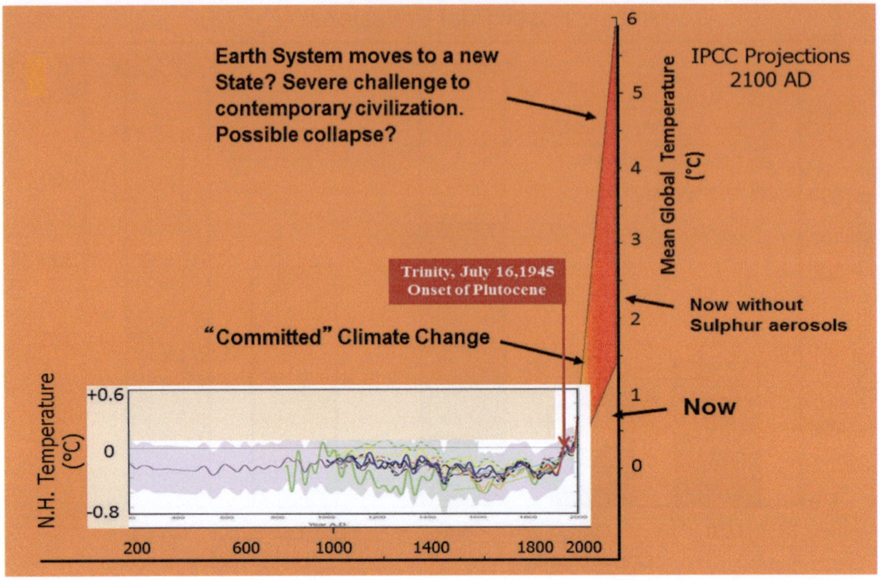

Fig. 1.20 The shift in state of the Earth climate and the onset of the Plutocene. (After Will Steffen, by permission)

Berger and Loutre (2002) (Fig. 1.21) suggest current climate change is leading to an exceptionally long interglacial period ahead, stating: "*The present day CO_2 concentration (397 ppm by 2012) is already well above typical interglacial values of ~290 ppmv. This study models increases to up to 750 ppmv over the next 200 years, returning to natural levels by 1000 years. The results suggest that, under very small insolation variations, there is a threshold value of CO_2 above which the Greenland ice sheet disappears. The climate system may take 50,000 years to assimilate the impacts of human activities during the early third millennium. In this case an 'irreversible greenhouse effect could become the most likely future climate. If the Greenland and west Antarctic ice sheets disappear completely, then today's 'Anthropocene' may only be a transition between the Quaternary and the next geological period.*"

The combined effects of sea level rise and extreme storms, flooding low-lying river valley systems (Ganges, Mekong, Yellow River, Nile, Indus, Tigris and Euphrates, Mississippi, Rhine, Thames), would inundate major agricultural regions and agrarian populations, curtailing food supplies to urban centers. The stress on global networks of food, fuel, technology and medication may prove calamitous. Depending on how far global warming proceeds and the extent to which humanity's nuclear arsenal is released, by accident or design, some members of the species may survive, mostly those genetically adapted to extreme conditions in remote parts of the globe. Arctic temperatures will still allow human body temperature comfort zones and the intensified hydrological cycle will enhance precipitation in the Siberian Taiga, possibly benefiting Caribou herds and nomad Laps. Depending on

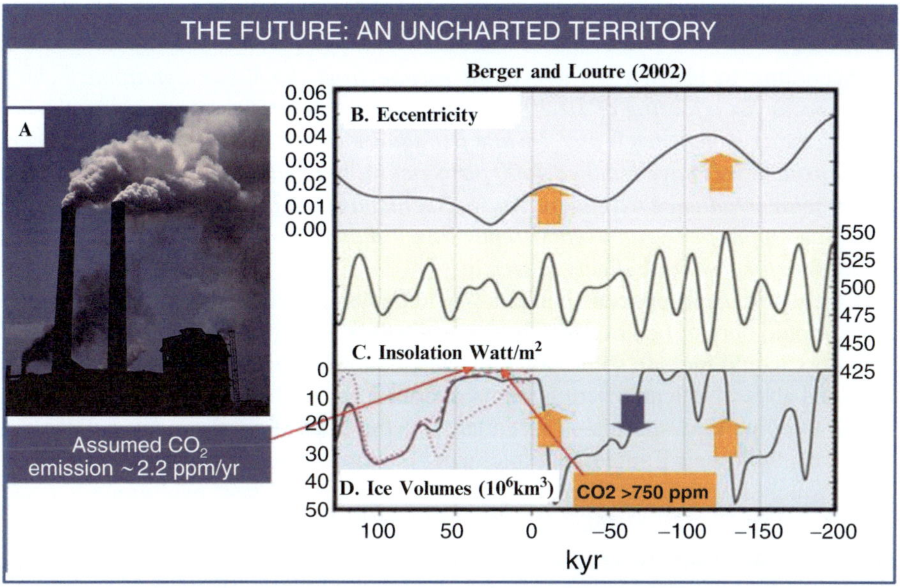

Fig. 1.21 (**a**) Smokestacks from a wartime production plant, World War II (https://commons.wikimedia.org/wiki/File:AlfedPalmersmokestacks.jpg); (B,C,D) An exceptionally long interglacial ahead?: Long-term variations of eccentricity (**b**), June insolation at 65°N (**c**), and simulated Northern Hemisphere ice volume (**d**) for – 200,000 years before the present to +130,000 from the present. For the future, three CO2 scenarios were used: last glacial-interglacial values (*solid line*), a human-induced concentration of 750 ppmv (*dashed line*), and a constant concentration of 210 ppmv (*dotted line*). Berger and Loutre (2002) (AAAS by permission)

the survival of land and marine life, small human clans like the Inuit may survive, provided ocean acidification and intense radioactivity do not destroy the base of the food chain. Humans may also survive in volcanic ocean islands well above sea levels and in well-watered sheltered mountain valleys. In such areas buffering of rising temperatures by increased clouding and heavier precipitation may allow a habitat for indigenous humans whose genetic characteristics, skin and respiratory system have been adapted to extreme humidity and heat over pre-historic periods, for example in the foothills of the Himalayas. However, rising temperatures would lead to burning of many forest regions by bush fires during dry seasons. Sea level rise and encroaching desert conditions, such as the advance of Sahara-like environment into southern Europe, would drive populations northward.

With emissions from fossil fuels and cement tracking toward 570 GtC (Global Carbon Budget 2016[46]), global fossil fuel reserves exceeding >2000 GtC, recoverable resources exceeding 11,000 GtC and unconventional resources such as coal seam gas, oil shale and tar sands constituting open ended resources (Fig. 1.19), the Earth's biosphere is in danger. Where one Gigaton Carbon raises atmospheric CO_2 by ~0.5 ppm, greenhouse gas concentrations exceeding 1000 ppm CO_2 may result

[46] http://www.earth-syst-sci-data.net/7/47/2015/essd-7-47-2015.pdf

in a runaway phenomenon (MIT 2012), referred to by Hansen et al. as a Venus syndrome[47,48]

According to Hansen[49] "*there is no escape from the Venus Syndrome, which could occur for a forcing of 10-20 watts per square meter. For comparison, the net forcing today is between 0 and 3 watts per square meter. Although in the past carbon dioxide levels have reached 4000 parts per million without a runaway warming effect, solar irradiance was lower. And today humans are increasing carbon dioxide levels at 2 ppm per year, 10,000 times faster than natural rates, which does not allow time for feedback effects to kick in.*"

Likely post-Anthropocene habitats would be dominated by grasses. Much of the forest would burn or be flooded. The grasses will be populated by arthropods. Other survivors would include some polar species, other heat and radiation tolerant species, and abyssal volcanic vent habitats of crustacean and extremophile bacteria. In so far as some bird species—descendants of the fated dinosaurs may eclipse the mammals with their limited heat tolerance, the circle would be complete. Vacated habitat niches would be filled by surviving species—a new evolutionary cycle would commence. Given detailed studies of past climates, paleontology, evolutionary biology and the nature of mass extinctions of species, clues for the aftermath of the ongoing sixth mass extinction of species are multiplying (Bradshaw and Brook 2006, Kolbert 2007, Lynas 2008, Chandler 2011) (Chapters 3 and 4).

1.3 Paradise Lost

In 2009 Joachim Hans Schellnhuber, Director of the Potsdam Climate Impacts Institute and Climate Advisor to the German Government, stated: "*We're simply talking about the very life support system of this planet*".[50] The consequences for the biosphere of accelerating climate change are discussed by Baronsky (2012) in the following terms: "*Localized ecological systems are known to shift abruptly and irreversibly from one state to another when they are forced across critical thresholds. Here we review evidence that the global ecosystem as a whole can react in the same way and is approaching a planetary-scale critical transition as a result of human influence.*" and "*Climates found at present on 10–48 % of the planet are projected to disappear within a century, and climates that contemporary organisms have never experienced are likely to cover 12–39 % of Earth. The mean global temperature by 2070 (or possibly a few decades earlier) will be higher than it has been since the human species evolved.*" The devastation of habitats amounting to a crisis in the biosphere has become manifest (Vitousek 1994), including a rapid polar-ward

[47] https://bravenewclimate.com/2010/05/09/clarons-despair/

[48] https://www.amazon.com/Storms-My-Grandchildren-Catastrophe-Humanity/dp/1608195023

[49] https://bravenewclimate.com/2010/05/09/clarons-despair/

[50] http://www.reuters.com/article/us-climate-science-idUSTRE58R3UI20090928

shift of climate zones which affects the distribution of flora and fauna. According to Xu et al. (2013) and NASA (2013)[51] vegetation growth at Earth's northern latitudes increasingly resembles lusher latitudes to the south and temperature and vegetation growth at northern latitudes now resemble those found 4°–6° of latitude farther south. Arctic sea ice and the duration of snow cover are diminishing, the growing season is getting longer and plants are growing to a greater extent. In the Arctic and boreal areas the characteristics of the seasons are changing, leading to great disruptions for plants and related ecosystems. The Arctic's greenness is visible on the ground as an increasing abundance of tall shrubs and trees in locations all over the circumpolar Arctic. Greening in the adjacent boreal areas is more pronounced in Eurasia than in North America, driven by amplified greenhouse effects. Plants and organisms display remarkable adaptation and regeneration powers where medium to long-term environmental changes occur. However, abrupt transitions and tipping points in the physical environment may exceed the adaptive capacity of some species, resulting in extinction (Figs. 1.22 and 1.23) by 1950 near to 70% of Mediterranean and temperate forests and near 50% of tropical and sub-tropical forests were lost. Projected changes in biodiversity need to discriminate between the effects of climate and land-use change effects (de Chazal and Rounsevella 2009). Habitat loss leading to decreased species richness is the most common land-use change. Habitats most affected by climate change factors include grassland, shrubland, boreal forests, cool conifer forests and Tundra, whereas habitats mostly affected by land clearing are tropical forests, warm mixed forests and temperate deciduous forests. Threatened and lost mammal species concomitant with loss of forest habitats, particularly pronounced in tropical and sub-tropical habitats, are documented in reports by the Millennium Ecosystem Assessment program. In the oceans the marine food chain may collapse in part under the pressures of decreasing pH and increasing temperatures.

Coral reefs, including Australia's Great Barrier Reef (Figs. 1.24 and 1.25) and Carribean reefs (Morelle 2016) are particularly vulnerable to water acidification and to thermal bleaching of their photosynthesizing zooxanthellae. According to Veron (2009) "*if CO_2 levels are allowed to increase to 650–700 ppm, as is projected to occur later this century, a return to pre-industrial level will take a period of about 30,000 and 35,000 years. Initially acidification is buffered by bicarbonate–carbonate ion exchange, but once the buffers are overwhelmed the pH changes abruptly. The oceans will remain acidified until neutralized by the dissolving of marine carbonate rocks and the weathering of rocks on land, a hugely protracted process. When CO_2 levels increase to 560 ppm, the Southern Ocean surface waters will be undersaturated with respect to aragonite, and the pH will be reduced by about 0.24 units—from almost 8.2 today to a little more than 7.9. At the present rate of acidification, all reef waters will have a Ω aragonite of 3.5 or less by the middle of this century. Should CO_2 levels reach 800 ppm later this century, the decrease will be 0.4 units and dissolved carbonate ion concentration will have decreased by almost*

[51] http://www.nasa.gov/home/hqnews/2013/mar/HQ_13-069_Northern_Growing_Seasons.html

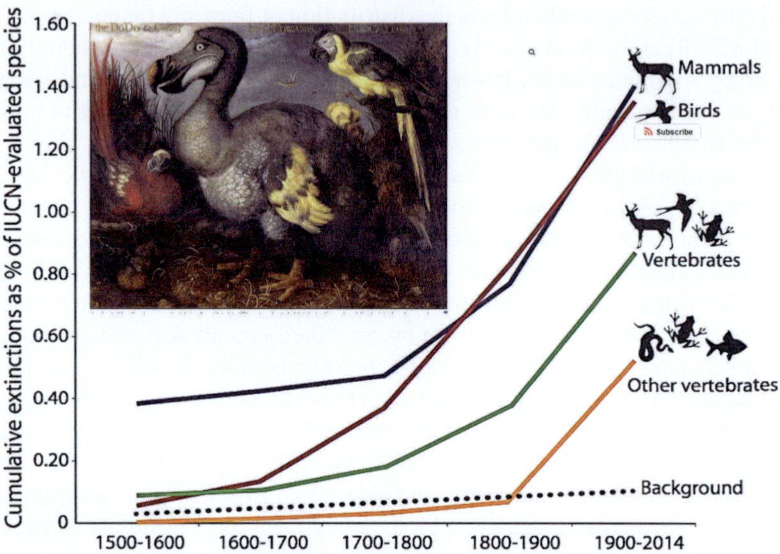

Fig. 1.22 Cumulative vertebrate species recorded as extinct in the wild by the IUCN (2012). *Dashed black line* represents background rate (http://advances.sciencemag.org/content/1/5/e1400253); *Inset:* The dodo, a flightless bird native to Mauritius, became extinct during the mid- to late seventeenth century due to habitat destruction and predation by introduced mammals (https://en.wikipedia.org/wiki/Holocene_extinction#/media/File:Edwards%27_Dodo.jpg)

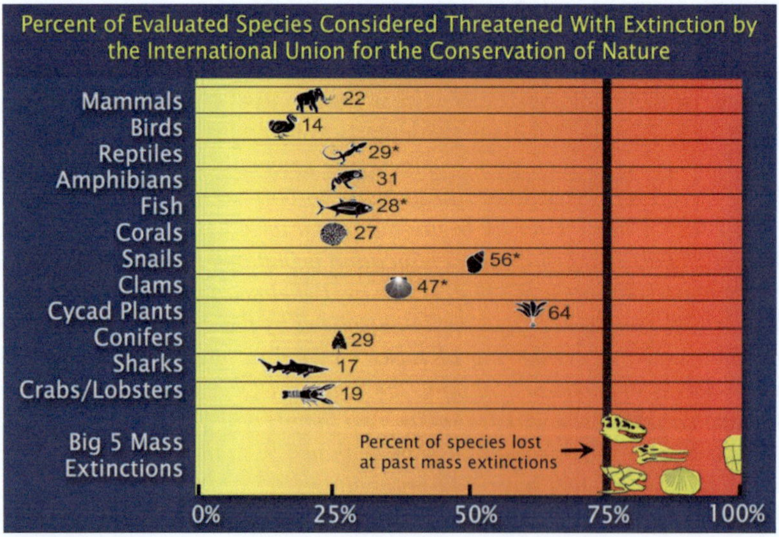

Fig. 1.23 Percent of species considered threatened with extinction by the International Union for the conservation of nature (Barnosky 2015) (http://www.huffingtonpost.com/anthony-d-barnosky/preventing-the-sixth-mass_b_6161284.html) (permission by A. Barnosky)

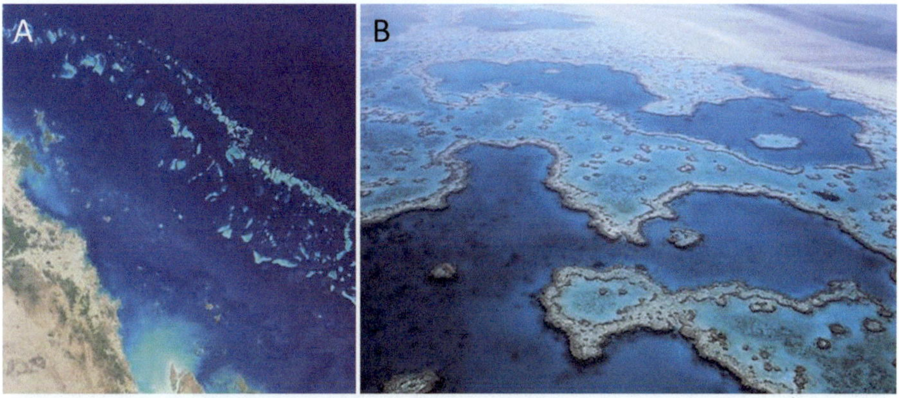

Fig. 1.24 The great barrier reef. Earth Observing System (EOS); (**a**) Satellite image (https://commons.wikimedia.org/wiki/File:GreatBarrierReef-EO.JPG); (**b**) Aerial photograph (https://pixabay.com/en/great-barrier-reef-coral-australia-527987/)

60%. At that point all the reefs of the world will be eroding relicts. The levels of CO_2 and pH predicted by the end of this century may not have occurred since the Middle Eocene, but the all-important rate of change we are currently experiencing has no known precedent. There can be no evolutionary solution for such a rate of change. Ultimately—and here we are looking at centuries rather than millennia—the ocean pH will drop to a point at which a host of other chemical changes, including anoxia, would be expected. If this happens, the state of the oceans at the end of K/T, or something like it, will become a reality and the Earth will enter the sixth mass extinction. Another 1–3 decades like our last will see the Earth committed to a trajectory from which there will be no escape."

In the oceans temperature-induced mass coral bleaching causing mortality on a wide geographic scale started when atmospheric CO_2 levels exceeded ~320 ppm (Veron 2009), triggering mass bleaching, mostly associated with El-Nino events world-wide. At current levels of more than 400 ppm reefs are committed to irreversible decline, exacerbated by the effects of increased severe weather events. The progressive onset of ocean acidification reduces coral growth and growth of coralline algae. With further rise toward 500 ppm CO_2 + methane terminal reef decline due to mass-bleaching would occur and shallow reef communities and biodiversity will be markedly reduced and face extinction (Hoegh-Guldberg 2007), with major effects on marine life and humanity. Should CO_2 levels reach 600 ppm surviving biota would be restricted to refuges, following the path of great mass extinctions of the past (Veron 2008, 2009, Ward 2015).

Fig. 1.25 (**a**) Corals Great Barrier Reef. Creative commons (https://en.wikipedia.org/wiki/File:Coral_Outcrop_Flynn_Reef.jpg; https://en.wikipedia.org/wiki/Coral_reef_protection), (http://earthfirstjournal.org/newswire/2014/07/02/caribbean-coral-reefs-could-vanish-in-20-years/) (Creative Commons); (**b**) Bleached corals of the Great Barrier Reef (https://commons.wikimedia.org/wiki/File:Bleachedcoral.jpg)

References

Abbott AN et al (2015) Constraints on ocean circulation at the Paleocene–Eocene thermal maximum from neodymium isotopes. Clim Past 12:837–847

Adler J (2013) Why fire makes us human. Smithsonian.com. http://www.smithsonianmag.com/science-nature/why-fire-makes-us-human-72989884/

Alvarez LW, Alvarez W, Asaro F, Michel HV (1980) Extra-terrestrial cause for the cretaceous-tertiary extinction: experimental results and theoretical interpretation. Science 208:1095–1085

Barnosky AD (2015) Approaching a state shift in Earth's biosphere. Nature 486:52–58

Beerling DJ, Lomax BH, Royer DL, Upchurch GR, Kump LR (2002) An atmospheric pCO2 reconstruction across the cretaceous-tertiary boundary from leaf mega fossils. Proc Natl Acad Sci U S A 99(12):7836–7840

Beerling DJ, Royer D (2011) Convergent Cenozoic CO 2 history. Nat Geosci 4:418–420

Berger A, Loutre MF (2002) An exceptionally long interglacial ahead? Science 297:1287–1288

Berner RA (2004) A model for calcium, magnesium and sulfate in seawater over Phanerozoic time. Am J Sci 304:438–453

Berner RA (2006) GEOCARBSULF: a combined model for Phanerozoic atmospheric O2 and CO2. Geochim Cosmochim Acta 70:5653–5664

Berner RA, Vanderbrook JM, Ward PD (2007) Oxygen and evolution. Science 316:557–558

Bowman DM et al (2009) Fire in the Earth system. Science 324:481–484

Braun H, Ditlevsen P, Chialvo DR (2008) Solar forced Dansgaard-Oeschger events and their phase relation with solar proxies. Geophys Res Lett 35(6):L06703–L06802

Bradshaw CJA, Brook BW (2006) Life and death on Earth – the Cronus hypothesis. J Cosmol 2:201–209

Broecker WS (2000) Abrupt climate change: causal constraints provided by the paleoclimate record. Earth Sci Rev 51:137–154

Broecker WS (2006) Abrupt climate change revisited. Glob Planet Chang 54:211–215

Broecker WS, Stocker TF (2006) The Holocene CO2 rise: anthropogenic or natural? Eos 87(3):27

Bowen GJ et al (2015) Two massive, rapid releases of carbon during the onset of the Paleocene–Eocene thermal maximum. Nat Geosci 8:44–47

Carlson AE (2010) What caused the younger dryas cold event? Geology 38:383–384

CDIAC (Carbon Dioxide Information Analysis Centre) (2017) 800,000-year Ice-Core records of atmospheric carbon dioxide (CO2). US Department of Energy, Oak Ridge http://cdiac.ornl.gov/trends/co2/ice_core_co2.html

Ceballos G et al (2015) Accelerated modern human–induced species losses: entering the sixth mass extinction. Sci Adv 1(5):e1400253

Chandler J (2011) Feeling the heat. Melbourne University Press, Melbourne, p 292

Chandler M et al (1994) Joint investigations of the middle Pliocene climate II: GISS GCM Northern Hemisphere results. Glob Planet Chang 9:197–219

Cortese G, Abelmann A, Gersonde A (2007) The last five glacial-interglacial transitions: a high resolution 450,000-year record from the subantarctic Atlantic. Paléo 22:PA4203

Cui Y, Kump LR, Ridgwell AJ, Charles AJ, Junium CK, Diefendorf AF, Freeman KH, Urban NM, Harding IC (2011) Slow release of fossil carbon during the Palaeocene–Eocene thermal maximum. Nat Geosci 4:481–485

De Chazal J, Rounsevell MDA (2009) Land-use and climate change within assessments of biodiversity change: a review. Glob Environ Chang 19:306–315

deMenocal PB (2004) African climate change and faunal evolution during the Pliocene-Pleistocene. Earth Planet Sci Lett 220:3–24

Eby N, Zickfeld K, Montenegro A, Archer D (2009) Lifetime of anthropogenic climate change millennial time scales of potential CO 2 and surface temperature perturbations. J Clim 22:2501–2511

Espanol M et al (2016) Spatial and temporal Antarctic Ice Sheet mass trends, glacio-isostatic adjustment, and surface processes from a joint inversion of satellite altimeter, gravity, and GPS data. J Geophys Res 121:182–200
Foster GI, Rohling EJ (2013) Relationship between sea level and climate forcing by CO2 on geological timescales. Proc Natl Acad Sci 110:1209–1214
France-Lanford C, Derry LA (1997) Organic carbon burial forcing of the carbon cycle from Himalayan erosion. Nature 390:65–67
Ganopolski A, Rahmstorf S (2002) Abrupt glacial climate changes due to stochastic resonance. Phys Rev Lett 88:3–6
Ganopolski A, Winkelmann R, Schellnhuber HJ (2016) Critical insolation–CO2 relation for diagnosing past and future glacial inception. Nature 529:200–203
Garzione CN (2008) Surface uplift of Tibet and Cenozoic global cooling. Geology 36:1003–1004
Glikson AY (2013a) Fire and human evolution: the deep-time blueprints of the Anthropocene. Anthropocene 3:89–92
Glikson AY (2013b) The asteroid impact connection of planetary evolution with special reference to large Precambrian and Australian impacts. Springer, Dordrecht, p 149
Glikson AY (2014) Evolution of the atmosphere, fire and the Anthropocene climate event horizon. Springer, Dordrecht, p 174
Glikson AY (2016) Cenozoic mean greenhouse gases and temperature changes with reference to the Anthropocene. Glob Chang Biol 22:3843–3858
Glikson AY, Groves C (2016) Climate, fire and human evolution: the deep time dimensions of the Anthropocene. Springer, Dordrecht, p 226
Global Carbon Project (2016) Global Methane Budget. http://www.globalcarbonproject.org/methanebudget/
Gradstein FM, Ogg JG (2004) Geological time scale – why, how and where next. Taylor Fr Lethaia 37:175–181
Grey K, Walter R, Calver CR (2003) Neoproterozoic biotic diversification: snowball Earth or aftermath of the Acraman impact? Geology 31:459–462
Hansen JE (2007) Scientific reticence and sea level rise. Environ Res Lett 2:024002
Hansen JE (2012) Climatologist James Hansen on "Cowards in Our Democracies". Think Progress https://thinkprogress.org/climatologist-james-hansen-on-cowards-in-our-democracies-c197139f2df9#.1wxe1ict8
Hansen JE et al (2007) Climate change and trace gases. Phil Trans R Soc A 365A:1925–1954
Hansen JE et al (2008) Target atmospheric CO2: where should humanity aim? Open Atmos Sci J 2:217–231
Hansen JE, Sato M (2012) Paleoclimate implications for human-made climate change. Clim Chang 2012:21–47
Hansen JE et al (2011) Earth's energy imbalance and implications. Atmos Chem Phys 11:13421–13449. doi:10.5194/acp-11-13421-2011
Hansen JE et al (2012) Perception of climate change. http://www.pnas.org/content/109/37/E2415.abstract
Hansen JE et al (2016) Ice melt, sea level rise and superstorms: evidence from paleoclimate data, climate modeling, and modern observations that 2C global warming could be dangerous. Atmos Chem Phys 16:3761–3812
Hoegh-Guldberg O (2007) Coral Reefs under rapid climate change and ocean acidification. Science 5857:1737–1742
Hudson SR (2011) Estimating the global radiative impact of the sea-ice-albedo feedback in the Arctic. J Geophys Res 116:D16102
MIT Technology Review (2012) How likely is a runaway Greenhouse Effect on Earth? https://www.technologyreview.com/s/426608/how-likely-is-a-runaway-greenhouse-effect-on-earth/
IPCC (2016) Working group I: the scientific basis. http://www.ipcc.ch/index.htm
IPCC (2007a) Climate change 2007: working Group I: the physical science basis. https://www.ipcc.ch/publications_and_data/ar4/wg1/en/ch6s6-3-2.html
IPCC (2007b) Climate change 2007: working Group II: impacts, adaptation and vulnerability. https://www.ipcc.ch/publications_and_data/ar4/wg2/en/ch4s4-4-10.html

Kasting JF, Donahue TM (1980) The evolution of atmospheric ozone. J Geophys Res 85:3255–3263

Keller G (2005) Impacts volcanism and mass extinction: random coincidence or cause and effect? Aust J Earth Sci 52:725–757

Keller G et al. (2012) Volcanism, impacts and mass extinctions. Geoscientist, The Geological Society. https://www.geolsoc.org.uk/Geoscientist/Archive/November-2012/Volcanism-impacts-and-mass-extinctions-2

Klein RG, Edgar B (2002) The Dawn of human culture. Wiley, New York, p 288

Kolbert E (2007) Field notes from a catastrophe: man, nature, and climate change. Bloomsbury, London, p 279

Kompanichenko V (2000) Average lifetime of an intelligent civilization estimated on its global cycle. SAO/NASA Astrophysics Data System, 213, p 438. http://adsabs.harvard.edu/full/2000ASPC..213..437K

Kregenow DA, Swenson ER (2002) The lung and carbon dioxide: implications for permissive and therapeutic hypercapnia. Eur Respir J 20:6–11, http://erj.ersjournals.com/content/20/1/6

Kump LP (2011) The last great global warming. Scientific American, July 2011, 58–61. https://www.scientificamerican.com/article/the-last-great-global-warming/

Kutzbach JE et al (2010) Climate model simulation of anthropogenic influence on greenhouse-induced climate change (early agriculture to modern): the role of ocean feedbacks. Clim Chang 99:351–381

Lamont Doherty Observatory M (2016) Ocean circulation implicated in past abrupt climate changes. https://www.sciencedaily.com/releases/2016/06/160630145000.htm

Le Quere C et al (2016) Global carbon budget 2016. Earth Syst Sci Data 8:605–649

Lenton TM, Held H, Kriegler E, Hall JW, Lucht W, Rahmstorf S, Schellnhuber HJ (2009) Tipping points in the Earth system. Proc Natl Acad Sci U S A 105:1786–1793

Lewis AR et al (2007) Major middle Miocene climate change and the extinction of tundra communities: evidence from the transantarctic Mountains. US Geol Surv Extended Abstract 135

Lunt DJ et al (2008) Closure of the Panama Seaway during the Pliocene: implications for climate and Northern Hemisphere glaciation. Clim Dyn 30:1–18

Lynas M (2008) Six degrees: our future on a hotter planet. Harper-Collins e-book, New York

Midgley M (1984) Biological and cultural evolution. The institute for cultural research, monograph series no. 20, London NW2 3BW. http://www.i-c-r.org.uk/publications/monographarchive/Monograph20.pdf

Morelle R (2016) Caribbean coral reefs' could vanish in 20 years. BBC News. http://www.bbc.com/news/science-environment-28113331

Mullen L (2004) Multiple Impacts. Astrobiology Magazine. http://www.astrobio.net/news-exclusive/multiple-impacts/

NASA/PenState (1991) Ancient Climate Events; Paleocene Eocene Thermal Maximum. https://www.e-education.psu.edu/earth103/node/639

NASA (2013) Amplified greenhouse effect shifts north's growing season. http://www.nasa.gov/home/hqnews/2013/mar/HQ_13-069_Northern_Growing_Seasons.html

NASA (2016) The Sunspot Cycle. https://solarscience.msfc.nasa.gov/SunspotCycle.shtml

National Academies Press (2017) National Academy of Sciences. Linkages between arctic warming and mid-latitude weather patterns, p 38. https://www.nap.edu/read/18727/chapter/2

NOAA (2016) Trends in atmospheric carbon dioxide. A global network for measurements of greenhouse gases in the atmosphere. https://www.esrl.noaa.gov/gmd/ccgg/trends/

Olivier JGJ et al (2014) Trends in global CO_2 emissions. PBL Netherlands Environmental Assessment Agency, The Hague, p 59

Pagani M et al (2006) Arctic hydrology during global warming at the Palaeocene/Eocene thermal maximum. Nature 442:671–675

Petit JR et al (1999) 420,000 years of climate and atmospheric history revealed by the Vostok deep Antarctic ice core. Nature 399:429–436

Rahmstorf S (2003) Timing of abrupt climate change: a precise clock. Geophys Res Lett 30(10):17.1–17.4

Rahmstorf SR, Coumou D (2011) Increase of extreme events in a warming world. Proc Natl Acad Sci U S A 108:17905–17909
Rahmstorf S et al (2015) Exceptional twentieth-century slowdown in Atlantic Ocean overturning circulation. Nat Clim Chang 5:475–480
Raymo ME (2009) PLIOMAX: Pliocene maximum sea level project. PAGES News 17(2). http://moraymo.us/wp-content/uploads/2014/04/2009_raymoetal.pdf
Raymo ME, Ruddiman WF (1992) Tectonic forcing of late Cenozoic climate. Nature 359:117–122
Raymo MR et al (2011) Departures from eustasy in Pliocene sea-level records. Nat Geosci 4:328–332
Renne PR et al (2013) Time scales of critical events around. The cretaceous-Paleogene boundary. Science 339:684–687
Rhiannon S (2016) Antarctic research into warm water causing rapid glacial ice melt. ABC News. http://www.abc.net.au/news/2016-12-17/search-for-source-of-warm-water-melting-tottenglacier/8129202
Rignot E et al (2011) Acceleration of the contribution of the Greenland and Antarctic ice sheets to sea level rise. Geophys Res Lett 38(5):L05503
Roe G (2006) In defense of Milankovitch. Geophys Res Lett 33:L24703
Rohling EJ et al (2014) Sea-level and deep-sea-temperature variability over the past 5.3 million years. Nature 508:477–482
Royer DL, Berner RA, Montañez I, Neil P, Tabor J, Beerling DJ (2004) CO 2 as a primary driver of Phanerozoic climate. GSA Today 14:3
Royer DL, Berner RA, Park J (2007) Climate sensitivity constrained by CO 2 concentrations over the past 420 million years. Nature 446:530–532
Royer DL, Berner RA, Beerling DJ (2001) Phanerozoic atmospheric CO change: evaluating geochemical and paleobiological approaches. Earth-Sci Rev 54:349–392
Ruddiman WF (1997) Tectonic uplift and climate change. Springer, New York, p 380
Ruddiman WF (2003) The anthropogenic greenhouse Era began thousands of years ago. Climate Change 61:261–293
Rubens CL, Figueira EI, Cunha L (1998) Contamination of the oceans by anthropogenic radionuclides. Quím Nova 21:73–77
Sagan C (1980) Cosmos. Random House, New York, p 396
Scott MR, Salter PF, Halverson JE (1983) Transport and deposition of plutonium in the ocean: evidence from Gulf of Mexico sediments. Earth Planet Sci Lett 63:202–222
Sepkoski JJ (1998) Rates of speciation in the fossil record. Phil Trans R Soc Lond B353:315–326
SETI Institute (2016) The Drake Equation: what do we need to know about to discover life in space? http://www.seti.org/drakeequation
Shakun JD, Clark PU, He F, Marcott SA, Mix AC, Liu Z, Otto-Bliesner B, Schmittner A, Bard E (2012) Global warming preceded by increasing carbon dioxide concentrations during the last deglaciation. Nature 484:49–55
Siegenthaler U et al (2005) Stable carbon cycle–climate relationship during the late Pleistocene. Science 310:1313–1317
Smith SJ et al (2011) Anthropogenic sulfur dioxide emissions:1850–2005. Atmos Chem Phys 11(1116):2011
Solanki SK (2002) Solar variability and climate change: is there a link? A&G 43(5):5.9–5.13
Solomon S, Plattner GK, Knutti R, Friedlingstein P (2009) Irreversible climate change due to carbon dioxide emissions. Proc Natl Acad Sci 106(6):1704–1709
Stanley SM (2016) Estimates of the magnitudes of major marine mass extinctions in earth history. Proc Natl Acad Sci 113:E6325–E6334
Steffen W, Crutzen J, McNeill JR (2007) The Anthropocoene: are humans now overwhelming the great forces of nature? Ambio 36(8):614–621
Steffen W et al (2016) Stratigraphic and Earth System approaches to defining the Anthropocene. Earth Future 4:324–345

Steffensen JP et al (2008) High-resolution Greenland ice core data show abrupt climate change happens in few years. Science 321:680–684

US Geol Surv (2016) PRISM, Pliocene Research, interpretation and Synoptic Mapping. http://geology.er.usgs.gov/egpsc/prism/

Vaughan A (2015) Humans creating sixth great extinction of animal species, say scientists. https://www.theguardian.com/environment/2015/jun/19/humans-creating-sixth-great-extinction-of-animal-species-say-scientists

Velicogna I (2009) Increasing rates of ice mass loss from the Greenland and Antarctic ice sheets revealed by GRACE. Geophys Res Lett 36(19):L19503

Velicogna I (2014) Regional acceleration in ice mass loss from Greenland and Antarctica using GRACE time-variable gravity data. Geophys Res Lett 41:8130–8137

Velicogna I, Wahr J (2005) Greenland mass balance from GRACE. Geophys Res Lett 32:L18505

Veron JEN (2008) Mass extinctions and ocean acidification: biological constraints on geological dilemmas. Coral Reefs 27:459–472

Veron JEN (2009) The coral reef crisis: the critical importance of <350 ppm CO2. Mar Pollut Bull 58:1428–1436

Vitousek PM (1994) Beyond global warming: Ecology and global change. Ecology 75(7):1861–1876

Vizcarra N (2014) Strange bedfellows: Is the delicate relationship between fire and plants changing? Earthdata. https://earthdata.nasa.gov/user-resources/sensing-our-planet/strange-bedfellows-2014

Ward P (2015) The Medea hypothesis: is life on earth ultimately self-destructive? Princeton University Press, Princeton, p 208

Wagner F, Aaby B, Visscher H (2002) Rapid atmospheric CO 2 changes associated with the 8,200-years-B.P. Cooling event. Proc Natl Acad Sci U S A 99:12011–12014

Wright JD, Schaller MF (2013) Evidence for a rapid release of carbon at the Paleocene-Eocene thermal maximum. Proc Natl Acad Sci 110(40):15908–15913

World Building (2016) How many survivors would be left after a global nuclear war? http://world-building.stackexchange.com/questions/28335/how-many-survivors-would-beleft-after-a-global-nuclear-war

Wynn G (2014) How much worse is a 4 degrees world? Climate Home. http://www.climatechangenews.com/2014/04/01/how-much-worse-is-a-4-degrees-world

Xu L et al (2013) Temperature and vegetation seasonality diminishment over northern lands. Nat Clim Chang 3:581–586

Yokoyama Y, Esat TM (2011) Global climate and sea level: enduring variability and rapid fluctuations over the past 150,000 years. Oceanography 24:54–69

Zachos J, Pagani M, Sloan L, Thomas E, Billups K (2001) Trends, rhythms, and aberrations in global climate 65 Ma to present. Science 292:686–693

Zachos J, Dickens GR, Zeebe RE (2008) An early Cenozoic perspective on greenhouse warming and carbon-cycle dynamics. Nature 451:279–283

Zeebe RE, Bada JL, Zachos JC, Dickens GR (2009) Carbon dioxide forcing alone insufficient to explain Paleocene–Eocene thermal maximum warming. Nat Geosci 2:576–580

Zeebe RE, Ridgwell A, Zachos JC (2016) Anthropogenic carbon release rate unprecedented during the past 66 million years. Nat Geosci 9:325–329

Chapter 2
The Breach in the Earth's Radiation Shield

The Moon's Eulogy

Exiled from mother Earth's womb
To outer space, the solar wind tomb
I orbit my Gondwana home
Where large green lizards used to roam

Now I watch white sheets of ice
Snaking glaciers' deep crevice
Mammoths graze the grassy steppes
From the tall trees descend some apes.

Vast floods encroach Mount Ararat
Sweeps a big arc, all its ports shut
Receding water bare rich soil
Adam's sons till harvest's toil

Slave armies march carting big stones
Erect colossus, the Gods to atone
Launch death orgies scorch the fields
Returning warriors on shields

Tumour celled cancer cities spreads
Poison water, the green fades
In burning skies vanished ozone
Red burning eyes, life's former zone

Missiles glow in orbit's decay
I shed a tear to Gaea's last day
Bidding adieu to Prometheus story
In surrender to Fates and betrayal of glory

'The splitting of the atom has changed everything
Save man's way of thinking
And thus we drift toward unparalleled catastrophes'.
Albert Einstein (http://investmentwatchblog.com/the-splitting-of-the-atom-changed-everything-save-mans-mode-of-thinking-thus-we-drift-towards-unparalleled-catastrophe-einstein/)

A.Y. Glikson, *The Plutocene: Blueprints for a Post-Anthropocene Greenhouse Earth*, Modern Approaches in Solid Earth Sciences 13,
DOI 10.1007/978-3-319-57237-6_2

Abstract The Earth's atmosphere and magnetic field constitute radiation shields which absorb and deflect high-energy radiation from the solar wind and cosmic rays, protecting the critical DNA and RNA biomolecules which form the basis of life from "genetic damage (Lomax 2013). The mining of Uranium, enrichment of Uranium-235 and the production of the artificial Plutonium-238 and 240 isotopes have opened a source of radioactive radiation released to the Earth's atmosphere, water and soil, as recorded as a distinct anthropogenic radioactive layer in the deep oceans. Nuclear tests in remote parts of the Earth have already commenced a nuclear war against indigenous people, including in Japan, Marshall Islands, Novaya Zemlya, Kazakhstan, Polynesia and the Sahara and Gobi Deserts. Despite the above, proponents of nuclear energy, assuming responsible human handling of fissile materials in the future, propose nuclear energy replaces carbon-based combustion systems. However, the fallibility of the nuclear industry has already been demonstrated by major nuclear accidents, such as in Chernobyl and Fukushima, as well as radioactive spills. Nuclear energy facilities have facilitated lateral proliferation of nuclear weapons. Extrapolating from ~500 megaton tests and release of plutonium from satellite accident, radiation levels from a 1000 and a 10,000 megaton war would reach a scale of 4.2×10^{11} Curies and 4.2×10^{12} Curies, respectively. Consequent proliferation of radiation hot spots around the world and triggered nuclear winter conditions arising from global smoke and soot released from burnt cities would result in a loss of life on the scale of more than one billion. The longer term consequences of a nuclear conflict are difficult to evaluate. Given the history of the atomic age, *H. sapiens* can hardly be trusted to limit the destructive potential of the splitting of the atom.

The splitting of the atom and harnessing of radioactive sources by *H. sapiens* have created powerful sources of ionizing radiation within the biosphere, threatening radiation levels adverse to animals and plants. The bombing of Hiroshima and Nagasaki, followed by the atomic and hydrogen bomb tests in the Marshall Islands, Novaya Zemlya, Moruroa Atoll, the Sahara, Lop Nur in Gobi Desert, Maralinga in South Australia, Monte Bello islands,[1] perpetrating nuclear experiments on indigenous people and on nature in remote parts of the world, can be regarded as preambles for events unleashed by thousands of nuclear weapons ready for the main events. By 2009 this included approximately 23,335 weapons with ~6400 MT (megatons) yield (Global Nuclear Arsenals 2009)[2] and by 2016 15,375 weapons[3] (Fig. 2.1). This arsenal, capable of destroying civilization and much of life on Earth hundreds of times,[4] keeps on being "modernized" in the name of "defense".

---

[1] https://www.ctbto.org/nuclear-testing/the-effects-of-nuclear-testing/the-soviet-unionsnuclear-testing-programme/

[2] http://www.nucleardarkness.org/include/nucleardarkness/files/global_nuclear_arsenal_in_number_2009.pdf

[3] http://www.ploughshares.org/world-nuclear-stockpile-report

[4] http://www.telesurtv.net/english/news/US-Has-Enough-Nuclear-Weapons-to-Destroy-Itself-4-Times-Over---20140914-0033.html

Fig. 2.1 (**a**) Over 2000 nuclear tests have been conducted in over a dozen different sites around the world. *Red* Russia/Soviet Union, *blue* France, *light blue* United States, *violet* Britain, *black* Israel, *yellow* China, *orange* India, *brown* Pakistan, *green* North Korea and *light green* (territories exposed to nuclear bombs). *Red*: Russia/Soviet Union; *blue*: France; *light blue*: United States; *purple*: United Kingdom. Diameter of circles are proportional to the magnitude of nuclear explosions (https://en.wikipedia.org/wiki/Nuclear_weapon#/media/File:Rael_Nuclear_use_locations_world_map.png); (**b**) A U.S. Air Force Boeing B-52H Stratofortress of the 2d Bomb Wing static display with weapons, at Barksdale Air Force Base, Louisiana (USA), in 2006. (https://commons.wikimedia.org/wiki/File:B-52H_static_display_arms_06.jpg); Numbers of nuclear weapons after 'Ploughshares' (http://www.ploughshares.org/world-nuclear-stockpile-report). See also Fig. 2.15

2.1 Ionizing Radiation and the Biosphere

The Earth's atmosphere and ionosphere constitute radiation screens absorbing, reflecting and deflecting high-energy radiation of the intense solar wind and Cosmic rays, the former accelerated by the high temperatures of the solar corona, or outer region of the Sun, to escape from the Sun's gravitational field. The particles include proton nuclei of hydrogen (89%), nuclei of helium (10%) and heavier nuclei (1%), whose collision with atoms in the upper atmosphere creates pion particles which

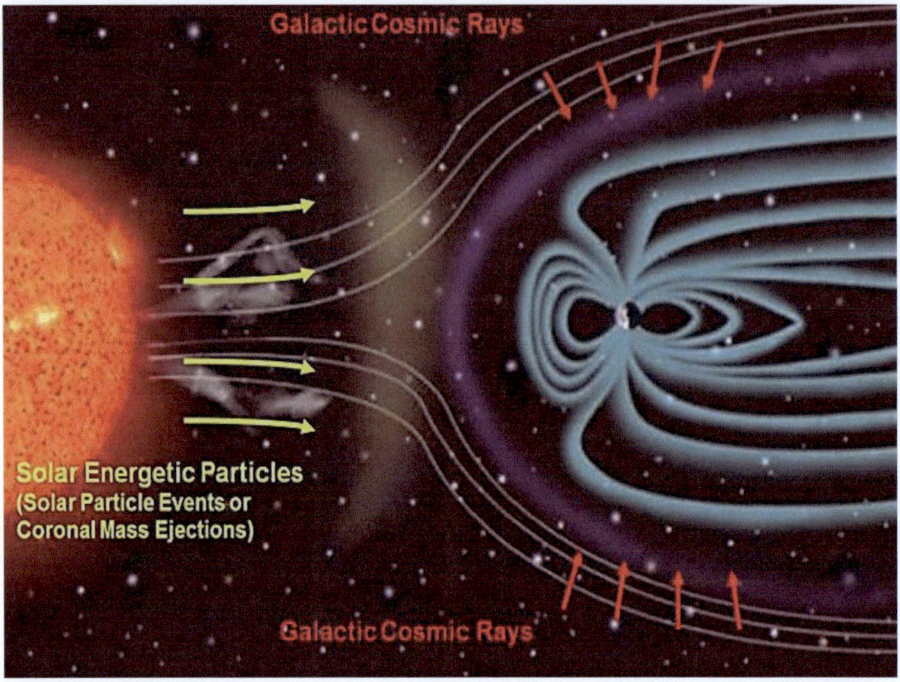

Fig. 2.2 Sources of ionizing radiation in interplanetary space. NASA JPL (http://photojournal.jpl.nasa.gov/catalog/PIA16938)

decay emitting muons.[5] Galactic cosmic rays originate from the *Milky Way* and extra-galactic cosmic rays from other galaxies.[6] A very small fraction consists of stable particles of antimatter, such as positrons or antiprotons.[7] Cosmic microwave radiation consists of low energy photons (2.78 Kev) and neutrinos representing relic primordial radiation, a distant echo of the Big Bang, mostly harmless to biomolecules (Fig. 2.2). The solar wind contains approximately 1–10 protons per cubic centimeter moving outward from the Sun at velocities of 350–700 km per-second, creating an ion flux of 10^8–10^9 ions per square centimeter per second. Energies range between 1.5 and 10 thousand electron volts (keV), responsible for creating the tail of Earth's magnetosphere and tails of comets. During solar flares, the proton velocity, flux, plasma temperature, and associated turbulence increase substantially.[8] The diversion of high energy particles of the solar wind by the Earth's magnetic field and the absorption of cosmic rays by oxygen and nitrogen molecules, and of

[5] https://home.cern/about/physics/cosmic-rays-particles-outer-space

[6] https://www.britannica.com/topic/cosmic-ray

[7] https://en.wikipedia.org/wiki/Cosmic_ray

[8] https://www.britannica.com/topic/solar-wind

ultraviolet radiation by the ozone layer, allow the survival of DNA-RNA biomolecules,[9] the basic elements of life.

Ionizing radiation is induced by the following particles[10]:

- Alpha particles[11] (Helium particles) consist of two protons and two neutrons produced when heavy elements decay to large high-energy particles propagated at a speed of 16,000 km/sec, about one tenth the speed of light, that can knock about 450,000 electrons off atoms. Alpha particles are absorbed by air and other media but are dangerous if inhaled or swallowed where they can disturb internal body tissues and organs. Radon gas, emitted from decay of ^{226}Ra, emits alpha particles. Alpha particles emitting ^{239}Pu have long half-life of up to 24,100 years.
- Beta rays[12] are small energetic electrons emitted from nucleus of unstable isotopes at a speed of 270,000 km/sec. The rays can be stopped by 3 m of air or a thin alumina sheet. Beta rays are capable of penetrating deep into living matter, damaging chemical links between living molecules of the cell or causing some permanent genetic change in the cell nucleus.
- Gamma Rays[13] (X-rays) constitute high-frequency electromagnetic ionizing radiation and consist of high-energy photons or light waves harmful to living cells through the transfer of energy to surrounding cells. They are produced by decay of atomic nuclei from a high energy state to a lower energy state, for example by atmospheric interactions with cosmic ray particles or lightning strikes.

The rounded total radiation from natural resources (3.10 millisievert/year)[14] and from artificial human-induced sources (radon, medical, internally inhaled radionuclides – 3.10 millisievert/year) total 6.2 millisievert/year.[15] The combination the solar wind and cosmic rays radiation penetrating the atmosphere induce radiation at the surface of 2.42 millisievert (cosmic 0.39 mSv; air – 1.26 mSv; terrestrial – 0.48 mSv; internal body due to food and other ingested components – 0.29 mSv).[16] The background radiation from cosmic rays increases with altitude, from ~0.3 mSv per year for sea-level areas to 1.0 mSv per year for higher altitudes. Airline crews flying long distance high-altitude routes can be exposed to an extra annual dose of 2.2 mSv (Tables 2.1 and 2.2).

[9] https://www.ncbi.nlm.nih.gov/pubmed/23849504 j

[10] http://www.oasisllc.com/abgx/radioactivity.htm.

[11] https://www.britannica.com/science/alpha-particle

[12] https://www.britannica.com/science/beta-decay

[13] https://www.britannica.com/science/gamma-ray

[14] Units of ionizing radiation are expressed in terms of absorbed radiation (Gray, Rad), emitted radiation (Curie, Becquerel) and calibrated biological effects of radiation (Sievert, Rem) The half-life values of radioactive isotopes range from a few days, i.e. Iodine (131I) 8.02 days, to tens of thousands of years i.e. Plutonium (239Pu) 24,100 years.

[15] http://www.physics.isu.edu/radinf/natural.htm, http://news.mit.edu/2011/explained-radioactivity-0328

[16] https://en.wikipedia.org/wiki/Cosmic_ray

Table 2.1 The world average and typical range of natural and artificial radiation (https://en.wikipedia.org/wiki/Cosmic_ray)

Radiation		UNSCEAR[a]	
		World	
Type	Source	Average mSv	Typical range mSv
Natural	Air	1.26	0.2–10.0
	Internal	0.29	0.2–1.0
	Terrestrial	0.48	0.3–1.0
	Cosmic	0.39	0.3–1.0
	Subtotal	2.4	1.0–13.0
Artificial	Medical	0.6	0.03–2.0
	Fallout	0.007	0–1+
	Others	0.0052	0–20
	Subtotal	0.6	0 to tens
Total		3	0 to tens

[a]United Nations Scientific Committee on the Effects of Atomic Radiation (UNSCEAR)

Table 2.2 Annual estimated average effective dose equivalent received by a member of the population of the United States (http://www.physics.isu.edu/radinf/natural.htm)

Source	NCRP[a] 160. 2006 mSv
Inhaled (Radon and Decay Products)	2.29
Other Internally Deposited Radionuclides	0.31
Terrestrial Radiation	0.19
Cosmic Radiation	0.27
Rounded total from natural source	3.1
Rounded total from artificial Sources (Medical, industrial, etc.)	3.1
Total	6.2

[a]National Council Radiation Protection (NCRP)

Radionuclides introduced into the marine environment undergo physical, chemical and biological processes related to the physical dispersion and chemical and biological interactions of the radionuclides with inorganic and organic matter, including bottom sediments and living organisms.[17] Radionuclides in the marine environment affect humans through ingestion of contaminated marine organisms, Cesium-137 being the most widely measured radionuclide. In the Irish Sea, the Baltic Sea and the Black Sea, concentrations of radionuclides depend on discharges from reprocessing facilities and from Chernobyl accident.

During 1945–1980 atmospheric and leaking underground nuclear tests (Johnston 2008)[18] released large quantities of anthropogenic radiogenic nuclides into the atmosphere and hydrosphere, subsequently deposited as fallout over the planetary surface (Hancock 2014). Short to medium half-life isotopic nuclides decay to

[17] http://www.scielo.br/scielo.php?script=sci_arttext&pid=S0100-40421998000100012

[18] http://www.johnstonsarchive.net/nuclear/tests/multimegtests.html

Table 2.3 Average latitudinal distribution of cumulative Pu-239, Pu-240 and Pu-238 fallout in mCi.km^2 (permission By Nature) (http://www.nature.com/nature/journal/v241/n5390/abs/241444a0.html)

	mCi per km^2			
			Pu-238	
Hemi-sphere	Lat. band°	Pu-239,240	Weapons	SNAP-9A
Northern	90–80	(0.10 ± 0.04)	(0.002 ± 0.001)	(<0.001)
	80–70	0.36 ± 0.05	0.009 ± 0.001	<0.001
	70–60	1.6 ± 1.0	0.038 ± 0.025	0.026 ± 0.015
	60–50	1.3 ± 0.2	0.031 ± 0.004	0.013 ± 0.004
	50–40	2.2 ± 0.5	0.053 ± 0.011	0.026 ± 0.011
	40–30	1.8 ± 0.6	0.042 ± 0.014	0.025 ± 0.015
	30–20	0.96 ± 0.07	0.023 ± 0.002	0.011 ± 0.004
	20–10	0.24 ± 0.10	0.006 ± 0.002	0.003 ± 0.002
	10–0	0.13 ± 0.06	0.003 ± 0.001	<0.001
Southern	0–10	0.30 ± 0.20	0.007 ± 0.005	0.010 ± 0.007
	10–20	0.18 ± 0.05	0.004 ± 0.001	0.036 ± 0.021
	20–30	0.39 ± 0.16	0.009 ± 0.004	0.070 ± 0.042
	30–40	0.40 ± 0.12	0.009 ± 0.003	0.061 ± 0.020
	40–50	0.35 ± 0.21	0.008 ± 0.005	0.069 ± 0.038
	50–60	(0.20 ± 0.09)	(0.005 ± 0.002)	(0.044 ± 0.023)
	60–70	(0.10 ± 0.04)	(0.002 ± 0.001)	(0.022 ± 0.012)
	70–80	(0.03 ± 0.01)	(0.001 ± 0.001)	(0.008 ± 0.005)
	80–90	(0.01 ± 0.004)	(<0.001)	(0.004 ± 0.002)

Results in parentheses were derived by extrapolation; error terms are standard deviations

negligible level, whereas longer-lived radionuclides persist for thousands of years. Plutonium isotopes ^{239}Pu (half-life 24,100 years) and ^{240}Pu (half-life 6561 years) constitute chronological markers in sediments, ice cores and coral cores. Pu inventory in soil will continue to supply Pu to sea-bottom sediments for millennia and centuries, depending on soil erosion, bioturbation rates and removal rates of dissolved Pu from the ocean.[19]

By 2009 the combination of nuclear tests and accidents has raised the radiation from fallout of Plutonium ^{239}Pu and ^{240}Pu to more than 325,000 ± 36.000 Curies (1.2 × 10^{16} Becquerel).[20] Of this 256,000 ± 33,000 Curies were deposited in the Northern Hemisphere and 69,000 ± 14,000 Curies in the Southern Hemisphere (Hardy et al. 1973) (Table 2.3).[21] Deposition of ^{238}Pu from nuclear tests and from an accident of a plutonium-fueled satellite (SNAP-9A) in 1964 amounted to deposition of a total of 21,600 curies, measured in soil samples.

[19] The release and persistence of radioactive anthropogenic nuclides (http://sp.lyellcollection.org/content/395/1/265.abstract).

[20] (1 Curie =37x109 Becquerel) (1 Becquerel = activity where one nucleus decays in a second). (Natural radiation level - 100 Bq/kg or 7000 Bq).

[21] http://www.nature.com/nature/journal/v241/n5390/abs/241444a0.html

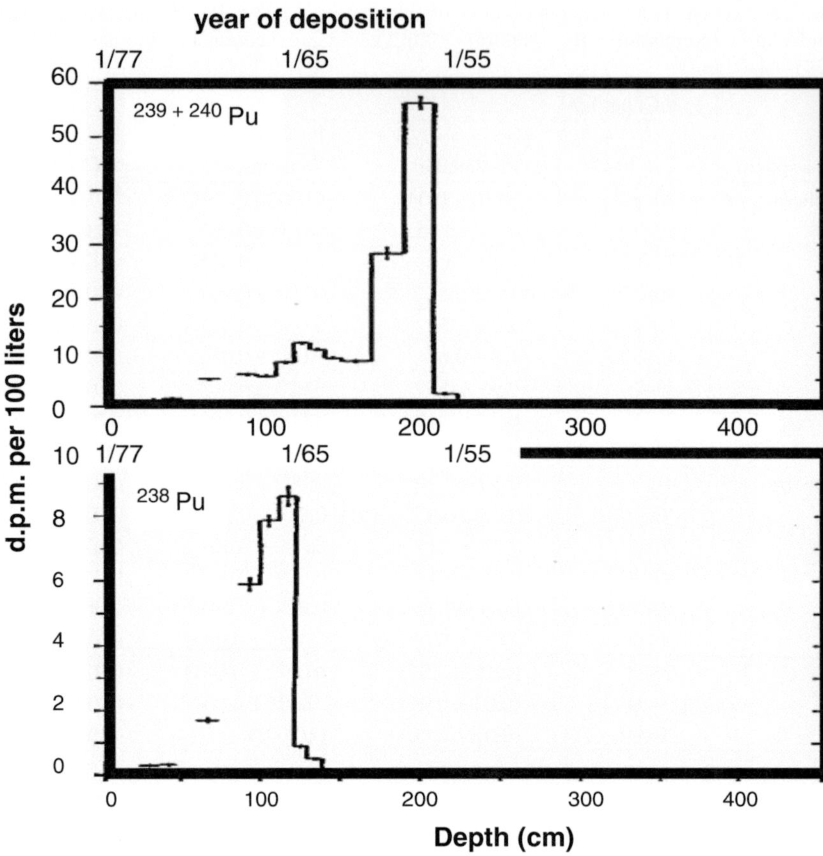

Fig. 2.3 ^{239}Pu + ^{240}Pu and ^{238}Pu activity in disintegrations per minute (DPM) at Dome C, Antarctica. *Horizontal lines* depict the depth of each layer sampled. *Vertical lines* within each sample represent uncertainties inherent in count statistics (Permission by Nature) (http://www.nature.com/nature/journal/v279/n5714/abs/279628a0.html)

Had the fallout from atmospheric nuclear tests and satellite accidents been distributed uniformly over the entire Earth surface (5.1×10^8 km^2; 5.1×10^{14} m^2), the amount of radiation from long-lived Pu isotopes would be approximately 6.37×10^{-4} Currie/km^2 or 2.35×10^7 Bq/km^2. Comparing the 2009 arsenal of 23,335 weapons, or 15,375 more recently, with a potential total yield of ~6400 megatons[22] with the 500 megatons exploded during 1945–1971 by the US and Soviet tests,[23] the current potential yield is at least an order of magnitude higher. In so far as a broad correlation can be assumed between explosive yield and radioactivity, the potential radiation of the modern arsenal would reach more than 3.10^6 Curies. A translation

[22] http://www.nucleardarkness.org/include/nucleardarkness/files/global_nuclear_arsenal_in_number_2009.pdf

[23] https://en.wikipedia.org/wiki/List_of_nuclear_weapons_tests#Soviet_Union

of the above figures to biological effects would depend on the geographic distribution, duration of radiation, cumulative effects and susceptibility.

Measurements of $^{239+240}Pu$ and ^{238}Pu fallout in Dome C on the high Antarctic plateau (Cutter et al. 1979) are indicated in Fig. 2.3. Peak radioactive depositions measured (disintegration per minute [dpm] in 100 liters snow) were reached in 1955 and 1965.

The effects of nuclear weapons are not limited to radiation.[24] A single nuclear weapon detonated above a city will ignite mass fire over an area of 50–500 km^2 from which living things have little chance to escape.[25] Vast areas of forests will go up in smoke and fires would burn for weeks in cities and industrial centers and vast stored fossil fuels. The fires will produce a thick smoke layer that will drastically reduce the amount of sunlight reaching the earth's surface. Darkness would persist for many weeks, rendering agricultural activity in the Northern Hemisphere virtually impossible if the war takes place during the growing season (Crutzen and Birks 1982). The release of dust, ash, particulate black carbon and sulphur dioxide from the fires would plunge large regions of the world into atmospheric conditions described as *nuclear winter* (Turco et al. 1983), threatening the survival of civilization and many species over large parts of Earth.

Despite a two-thirds reduction in global nuclear arsenals since 1986, the environmental consequences of nuclear war can still end human history as even a tiny fraction of the global nuclear arsenal exploded in large urban centers would result in catastrophic disruptions of the global climate and the protective stratospheric ozone layer. A nuclear war fought with several thousand weapons could even leave much of the Earth uninhabitable.[26]

By 2009 the world possessed a nuclear arsenal of Approximately 23,335 weapons with ~6400 MT (megatons) yield.[27] With time, the human warlike nature and the release of nuclear radiation by accident or design become probabilities, and eventually a certainty. Radioactive leaks from power generating plants and weapon manufacturing plants are reported in RADNET 2016,[28] for example leaks amounting to 0.74 million curies from Iodine-131 from the Hanford plant, Washington. Already the fallout products from atomic warfare, atomic tests, dumping at sea and accidents are manifest in soils and in deep ocean sediments and in ice. The develop-

[24] http://science.sciencemag.org/content/222/4630/1283

[25] http://www.nucleardarkness.org/include/nucleardarkness/files/global_nuclear_arsenal_in_number_2009.pdf

[26] Catastrophic Climatic Consequences of Nuclear Conflict. http://www.nucleardarkness.org/include/nucleardarkness/files/pdf/INESAP_2008_catastrophic_climatic_consequences_of_nuclear_conflict.pdf

[27] http://www.nucleardarkness.org/include/nucleardarkness/files/global_nuclear_arsenal_in_number_2009.pdf

[28] http://www.davistownmuseum.org/cbm/Rad14.html

ment of fleets of nuclear missiles, currently some 15,000 in strength[29,30,31,32] (Fig. 2.1), threatens regional to global radiation levels and a Nuclear Winter climate scenario[33] (Sect. I2.1). The build-up of radioactive reservoirs and waste products (Pearce 2015) threaten to return the Earth surface and oceans to transient radiative states such as existed prior to the development of the ozone layer[34] for periods on the scale of several 10^4 years.

Based on the brief ~2 million years episode a species mastering control over fire, a source of energy exceeding its own physiological capabilities by orders of magnitude, came to dominate the biosphere. Thus, whereas human respiration dissipates 2–10 calories per minute, a camp fire covering one square meter releases approximately 2×10^5 Calories/minute, magnifying the human-triggered energy output rate by more than 4 orders of magnitude. The output of a 1000 megawatt/hour power plant, expending some 2.5×10^9 calories/minute, surpasses the human energy release rate by more than 8 orders of magnitude.[35]

Future biospheres may be dominated by survivors of the genus Homo and by radiation-resistant arthropod species. Where the upper safe radiation limit for humans is about ~70 rad (700 Millisievert), irradiation data generated since the 1950s for over 300 arthropod species[36] indicate the dose needed for sterilization of arthropods varies from less than 500 rad for blaberid cockroaches to 30,000 rad or more for some arctiid and pyralid moths. Having originated about the Cambrian Explosion of life, when small bivalve-like shells were dated as 541–539 Ma in the classic outcrops of the Burgess Shale (Gould 1990), the highly advanced intelligence of insect colonies is likely to dominate future cycles of the terrestrial biosphere.

2.2 Late Anthropocene Nuclear Tests and Wars

Late Anthropocene nuclear wars and tests commenced with the dropping of atomic bombs on Hiroshima and Nagasaki at the end of World War II were followed by a long series of nuclear tests in remote parts of the world affecting indigenous people in the Pacific, the Arctic and parts of Africa and Asia. More than 500 megatons of nuclear yield were detonated in the atmosphere during 1945–1971 with a peak

---

[29] http://www.ploughshares.org/world-nuclear-stockpile-report

[30] http://www.icanw.org/the-facts/nuclear-arsenals/

[31] https://cnnd.crawford.anu.edu.au/sites/default/files/publication/cnnd_crawford_anu_edu_au/2015-02/printer_copy.pdf

[32] https://cnnd.crawford.anu.edu.au/sites/default/files/publication/cnnd_crawford_anu_edu_au/2015-02/printer_copy.pdf

[33] http://science.sciencemag.org/content/222/4630/1283

[34] http://history.nasa.gov/CP-2156/ch2.8.htm

[35] http://www.springer.com/gp/book/9783319225111

[36] https://www.ncbi.nlm.nih.gov/pmc/articles/PMC3507290/

during 1961–1962 when a total of 340 megatons atmospheric tests were conducted by the United States and Soviet Union.[37] The limited nuclear test ban treaty of 1963 ended atmospheric testing for the United States, Britain, and the Soviet Union, but France and China continued with nuclear testing (Simone et al. 1979).[38]

Cesium-137, a gamma emitter with a half-life of ~30 years, poses a serious threat as it parallels potassium chemistry and is taken into the blood and tissues of animals and humans. Strontium-90 is likewise dangerous as an electron emitter with a half-life of ~29 years following calcium chemistry and incorporated into the bones and teeth, particularly of young children. Iodine-131 with a half-life of only ~8 days is a similar threat to children due to its concentration in the thyroid gland. Plutonium-239, an alpha particles emitter, is incorporated in bones and may become lodged in the lungs, where its intense local radiation can cause cancer or other damage.[39]

A comprehensive compendium of nuclear events, accumulation and dispersion of long-lived radionuclides is reported at RADNET,[40] with major nuclear accidents recorded.[41] The issue of nuclear testing has been covered by numerous books, including *Radioactive Heaven and Earth* (1991).[42] Major accidents within the United States occurred in Rocky Flats, Colorado, Savannah River (South Carolina), Hanford Reservation (Washington), Oak Ridge Reservation (Idaho) and Fernald (Ohio), Los Alamos National Laboratory. Major accidents outside the US include Chernobyl, Fukushima, Tokaimura uranium processing plant, Sellafield (UK) and Dounreay (Scotland). The biggest bomb was the *Tsar Bomba*, a 50 megaton hydrogen bomb exploded on October 30, 1961 over Severny Island, Novaya Zemlya (McVicker 1954) (Fig. 2.6 and Fig. 2.7).[43]

Reports on nuclear testing by the US, Soviet Russia, UK, France and China are documented by the Preparatory Commission for the Comprehensive Nuclear Test Ban Treaty Organization (CTBTO 2012)[44,45]. Between the early 1940s and 1980 Earth received anthropogenic radioactive contamination from nuclear weapons testing with maximum releases in the mid-70s. Nuclear pollution from military uranium processing facilities has been widespread since the 1940s due to nuclear fission for military purposes (Aarkrog 1998; Meralli 2009), including (1) release of radionuclides to the Columbia River and the Pacific Ocean from reactors at Hanford,

---

[37] http://www.atomicarchive.com/Docs/Effects/wenw_chp2.shtml

[38] http://www.atomicarchive.com/Docs/Effects/wenw_chp2.shtml

[39] http://www.atomicarchive.com/Docs/Effects/wenw_chp2.shtml

[40] http://www.davistownmuseum.org/cbm/Rad14.html

[41] http://www.davistownmuseum.org/cbm/Rad16.html

[42] http://ieer.org/wp/wp-content/uploads/1991/06/RadioactiveHeavenEarth1991.pdf

[43] https://en.wikipedia.org/wiki/Tsar_Bomba

[44] The Comprehensive Nuclear-Test-Ban Treaty Organisation (CTBTO) https://www.ctbto.org/nuclear-testing/

[45] https://www.ctbto.org/nuclear-testing/

USA[46]; (2) from military nuclear facilities at Cheliabinsk (MAJAK)[47] which discharged fission products to the nearby Techa River of the Ob river system flowing to the Arctic Ocean; (3) Atmospheric testing of nuclear weapons peaked in the early 1960s as has radioactive contamination of the world's oceans; (4) In the mid-1970s ^{137}Cs contaminated waste from the nuclear reprocessing plant Sellafield in the UK could be traced all over the Northeastern Atlantic Ocean.[48] (4) In 1986 the Chernobyl accident[49] contaminated the Baltic and the Black Seas. In the 1990s military dumping activities were identified in the Arctic Ocean by the IAEA's IASAP (International Atomic Energy Agency.

Data regarding Anthropogenic Radionuclides in the Marine Environment: A Selected Bibliography are published by NOAA,[50] with summaries appended below (Boxes 2.1, 2.2, 2.3, 2.4).

Boxes 2.1, 2.2, 2.3, 2.4. Specifications of Nuclear Weapons and Their Effects[51] (Permission by Rod Nave, Hyperphysics 2016)

Box 2.1

Uranium Bomb

Using the energy release from the nuclear fission of uranium-235, an explosive device can be made by simply positioning two masses of U-235 so that they can be forced together quickly enough to form a critical mass and a rapid, uncontrolled fission chain reaction. That is not to say that this is an easy task to accomplish. First you must obtain enough uranium which is highly enriched to over 90% U-235, whereas natural uranium is only 0.7% U-235. This enrichment is an exceptionally difficult task, a fact that has helped control the proliferation of nuclear weapons. Once the required mass is obtained, it must be kept in two or more pieces until the moment of detonation. Then the pieces must be forced together quickly and in such a geometry that the generation time for fission is extremely short. This leads to an almost instantaneous buildup of the chain reaction, creating a powerful explosion before the pieces can fly apart. Two hemispheres which are explosively forced into contact can produce a bomb such as the one detonated at Hiroshima.

[46] http://www.alternet.org/environment/catastrophic-leak-found-hanford-nuclear-site-washington-state

[47] https://sometimes-interesting.com/2011/07/15/the-most-contaminated-place-on-earth-chelyabinsk-40/

[48] https://www.newscientist.com/article/mg22530053-800-shocking-state-of-worlds-riskiest-nuclear-waste-site/

[49] http://www.world-nuclear.org/information-library/safety-and-security/safety-of-plants/chernobyl-accident.aspx

[50] http://www.lib.noaa.gov/researchtools/subjectguides/marine_radionuclides.pdf

[51] http://hyperphysics.phy-astr.gsu.edu/hbase/NucEne/bomb.html .

Box 2.2

Hydrogen Bomb

Because the thermonuclear explosive devices used hydrogen isotopes, (deuterium-tritium fusion), the resulting bombs were often called "hydrogen bombs". The first hydrogen bomb was detonated on November 1, 1952 at the small island Eniwetok in the Marshall Islands. It's yield was several megatons of TNT. The Soviet Union detonated a fusion bomb in the megaton range in August of 1953. The U.S. exploded a 15 megaton fusion bomb on March 1, 1954. It had a fireball 4.8 km in diameter and created a huge characteristic mushroom-shaped cloud. Analysis of the radioactive fallout from this bomb revealed it to be a fission-fusion-fission weapon, a "hydrogen bomb" with an outer sheath of natural uranium to increase the yield.

Box 2.3

Hiroshima

On August 6, 1945, a uranium fission bombwas detonated over the Japanese city of Hiroshima. The bomb, called "Little Boy" was a "gun-type" device which used an explosive charge to force two sub-critical masses of U-235 together. It was 28 inches in diameter and 120 inches long, a relatively small package to deliver an explosive force of some 20,000 tons of TNT by converting about 1 gram of matter into energy. This could be accomplished with a sphere of U-235 about the size of a baseball. This kind of device had never been tested, in contrast to the plutonium bomb which was dropped on Nagasaki three days later. No device like this has been used since, making the estimates of radiation exposure at Hiroshima very difficult. Casualties included both direct blast victims plus those who died from radiation-induced cancer in subsequent years.

The bomb was triggered to explode at a height of 550 meters (1800 ft), a height calculated to cause the widest area of damage.

In the detonation of the uranium fission bomb over Hiroshima, about 130,000 people were reported killed, injured, or missing. Another 177,000 were made homeless.

Box 2.4

Nagasaki

On August 9, 1945 a plutonium fission bomb was detonated over the Japanese city of Nagasaki, three days after a uranium fission bomb was dropped on Hiroshima. The bomb, called "Fat Man", was 128 inches long and had a diameter of 60.5 inches. It used implosion to compress the sub-critical assembly of plutonium. This kind of device had been tested less than a month before the drop, and was the subject of several other weapons tests after World War II. The explosive yield was about 20,000 tons of TNT, generated in about a microsecond.

The bomb was triggered to explode at a height of 550 meters (1800 ft), a height calculated to cause the widest area of damage.

Hiroshima-Nagasaki[52]: The atomic age was born on July 16, 1945 at Alamogordo Trinity test, New Mexico, followed by extensive tests in the western USA, mainly Nevada, with major radiation effects on western and central parts of the continent (Fig. 2.4).[53] The test was succeeded at 8:15 a.m., 6 August, 1945, by a 64 kg bomb[54] containing highly enriched ^{235}U, dropped over Hiroshima by the US plane Enola Gay, exploding about 580 m above ground and releasing 1.4 million curies, with consequent loss of 90,000–146,000 people from direct effects and radiation exposure (Box 2.2). A subsequent 4670 kg bomb containing a core of ^{239}Pu was dropped on 9 August on Nagasaki, killing 39,000–80,000 people (Box 2.3). Video documentation of both bombings has been produced at the time[55] (Fig. 2.5).

---

[52] http://www.livescience.com/45509-hiroshima-nagasaki-atomic-bomb.html

[53] https://en.wikipedia.org/wiki/Environmental_radioactivity#/media/File:US_fallout_exposure.png

[54] https://en.wikipedia.org/wiki/Little_Boy

[55] http://www.atomicheritage.org/history/little-boy-and-fat-man

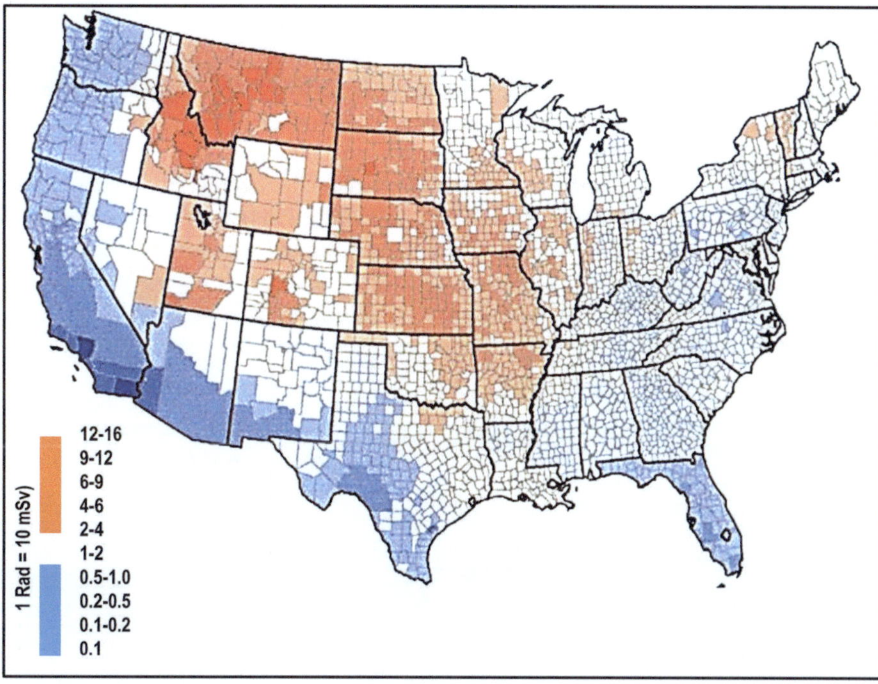

Fig. 2.4 Per-capita thyroid doses in the continental United States resulting from all exposure routed from all atmospheric nuclear tests conducted at the Nevada test site from 1951 to 1962 (https://en.wikipedia.org/wiki/Environmental_radioactivity#/media/File:US_fallout_exposure.png)

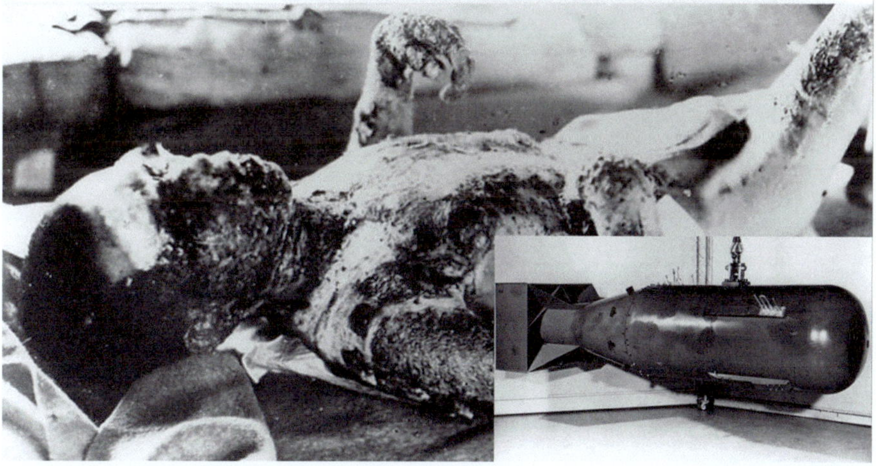

Fig. 2.5 A victim of the Hiroshima atom bomb (https://commons.wikimedia.org/wiki/File:Victim_of_Hiroshima_atomic_bombing_3.jpg); *Inset*: A mockup of Little Boy, the atomic bomb dropped over Hiroshima, Japan, on Aug. 6, 1945 (https://en.wikipedia.org/wiki/Little_Boy)

2.2.1 *The Soviet Union Nuclear Tests*[56]

Nuclear tests of 456 atomic and thermonuclear devices were conducted in the mountainous archipelago of Novaya Zemlya, between 1949 and 1989[57] (Fig. 2.6 and Fig. 2.7) and in Semipalatinsk, Kazakhstan (Miklós 2013; Ivanov 1996) (Fig. 2.8), and Five of the surface tests were unsuccessful and resulted in the dispersion of plutonium into the environment, contaminating villages of Nenets-people in Novaya-Zemlya. The Semipalatinsk's Degelen Mountain nuclear test facility was the largest underground nuclear test site in the world, consisting of 181 separate tunnels. Between October 1961 and October 1989, 224 tests were conducted. The approximate cumulative explosive yield of the tests conducted before 1963 was 6.4

Fig. 2.6 The Tsar Bomba hydrogen bomb, the most powerful nuclear weapon ever detonated with a yield of 57 megaton TNT, detonated at Severny Island, Novaya Zemlya (https://en.wikipedia.org/wiki/Tsar_Bomba#/media/File:Tsar_photo11.jpg); *Inset*: Yulii Borisovich Khariton (1904-1996), the USSR chief nuclear physicist (https://en.wikipedia.org/wiki/Yulii_Borisovich_Khariton#/media/File:Yulii_Borisovich_Khariton_1924.jpg)

[56] https://www.ctbto.org/nuclear-testing/the-effects-of-nuclear-testing/the-soviet-unionsnuclear-testing-programme/

[57] http://www1.american.edu/ted/ice/novalya.htm

Fig. 2.7 Andrei Dmitrievich Sakharov (21 May 1921 – 14 December 1989) – a Russian nuclear physicist, Soviet dissident, an activist for disarmament, peace and human rights (https://en.wikipedia.org/wiki/Andrei_Sakharov)

Fig. 2.8 Nuclear bomb crater, Semipalatinsk (https://commons.wikimedia.org/wiki/File:Crater_-_Flickr_-_The_Official_CTBTO_Photostream.jpg)

Megatons (Mt). Between 1945 and 1984 over 2000 kilos of plutonium have been released into the atmosphere.[58]

The Soviets conducted these tests without any regard for the effects on the local environment or the almost quarter-million inhabitants of the area. The biggest had a 1640 feet (500 m) diameter and 260 feet (80 m) maximum depth, formed in 1965. "That lake is void of any living creature. Children with genetic diseases, leukemia, infertility, and cancer are really common here. After the fall of the Soviet Union and the birth of Kazakhstan, in the summer of 1991, the site was closed. But a tenth of the country's total population – nearly 1.5 million people – have health problems. Lots of poor people still live in the most dangerous zone in a semi-nomadic way and sell the remaining scrap metal for money. One in every 20 children in the area is born with serious deformities, and half of them can't reach the age of 60. The drinking water from the nearest contaminated river has almost 100 times more tritium than the recommended limit."[59]

The total yield of the Novaya Zemlya tests is estimated at 273 Mt, about 94% of the total for all Soviet tests. Radioactive contamination extended to Alaska and northern Canada and Norway – located some 900 km away from Novaya Zemlya. Fallout from all past atmospheric weapons testing is still a major source of plutonium isotopes in the Arctic seas.[60] Large amounts of radioactive material associated with nuclear weapon testing had been dumped into the Barents and Kara seas. Novaya Zemlya was used as a graveyard for various nuclear weapons, submarines, and reactors sunk to the bottom of the ocean. As containers break down and submarines corrode, the materials inside have the potential to pollute marine life and disperse radioactivity into the ecosystem. Strong currents can then carry contaminants into fishing grounds and into the feeding areas of sea mammals and birds.

Novaya Zemlya was inhabited by nomadic peoples and reindeer before nuclear testing started. Most of the reindeer either died or were transported to the mainland. Roughly 500 people were relocated due to the testing program. They could not assimilate with their new surroundings and many succumbed to tuberculosis, syphilis, drunkenness, and suicide. An entire ethnic branch of the Nenets-people was completely wiped out. Research conducted by Bauer (2005) asserted that a significant association between solid cancer and radiation dose was found in the Semipalatinsk region.

In the Russian Arctic, increases in mortality rates in cancer, and in blood, skin, and oncological diseases were reported in the Arkhangelsk region. The Russian Academy of Medical Sciences observed the cancer death rate in Chukotka jumped from 10% of the population in 1970 to 27% in 1988. The Soviet Union's 26 indigenous peoples claimed the indigenous Inuit population was becoming ill from radioactive contamination in the environment. According to Peter Gizewski[61] scien-

[58] http://www1.american.edu/ted/ice/novalya.htm

[59] http://io9.gizmodo.com/5988266/the-tragic-story-of-the-semipalatinsk-nuclear-test-site

[60] https://www.ctbto.org/nuclear-testing/the-effects-of-nuclear-testing/the-soviet-unionsnuclear-testing-programme/page-4-effects-of-nuclear-weapon-testing-by-the-soviet-union/

[61] https://www.ctbto.org/nuclear-testing/the-effects-of-nuclear-testing/the-soviet-unionsnuclear-testing-programme/page-4-effects-of-nuclear-weapon-testing-by-the-soviet-union/

Fig. 2.9 Map of the East Urals Radioactive Trace, the area contaminated by the Kyshtym disaster (https://en.wikipedia.org/wiki/Kyshtym_disaster); *Inset*: Kyshtym memorial (https://en.wikipedia.org/wiki/Kyshtym_disaster)

tists noted that thousands of seals were dying off Russia's northern coastline as a result of radioactive pollution of the seabed.

A nuclear test conducted at the Totsk test site in the Arinbuk region of the southern Urals on 14 September 1954[62] continues to cause serious health and environmental effects decades later, with soil levels of plutonium-239 up to five times normal levels and high levels of cesium-137 contamination. The region's population suffers shorter life expectancy and a death rate 1.8 times higher than in other similar areas, a high infant mortality and a high rate of physical retardation in children (CTBTO[63]). (Fig. 2.8).

From 1948 radiation from the Mayak Kyshtym nuclear facility in the southern Urals (Rabi 2012), where Soviet atomic bombs were manufactured, affected some 17,245 workers. Radiation spills into the nearby river caused several breakouts of radiation sickness in villages downstream (Fig. 2.9). On the 29 September 1957

---

[62] http://www.albionmonitor.com/9611a/russia1950nuke.html

[63] https://www.ctbto.org/nuclear-testing/the-effects-of-nuclear-testing/the-soviet-unionsnuclear-testing-programme/

failure of the cooling system for a tank storing tens of thousands of tons of dissolved nuclear waste resulted in a chemical explosion with energy estimated at about 75 tons of TNT releasing 740 PBq (20 MCi) of fission products, of which 74 PBq (2 MCi) drifted off the site, creating a contaminated region of 15,000–20,000 km^2. An estimated 49–55 people died of radiation-induced cancer, 66 were diagnosed with chronic radiation syndrome, 10,000 people were evacuated from their homes, and 470,000 people were exposed to radiation[64,65] (Rabi 2012[66]).

Anthropogenic Radionuclides in the Arctic Ocean (Josefsson 1998[67]), originating in the atmospheric testing of thermonuclear bombs culminating in the early 1960s, led to global fallout estimated at a total effective dose of 3×10^7 Sievert disseminated by radioactive isotope ^{14}C. Atmospheric nuclear weapons tests account for 15% of the collective effective dose from all anthropogenic sources, and less than 5% of that from the natural sources. Global fallout at high latitudes was low, but currents from the North Atlantic have since deposition constituted a continuous transfer of radioactivity to the Nordic and Arctic Seas. Enormous quantities of decommissioned Russian nuclear reactors and radioactive waste were dumped into the Kara Sea in the Arctic Ocean north of Siberia over a course of decades, according to documents given to Norwegian officials by Russian authorities and published in Norwegian media.[68] Other significant sources of marine radioactivity are from plants reprocessing nuclear waste in northern Europe, whose marine discharges are transported with sea currents towards the Arctic regions. A possible threat also comes from potential sources such as radioactive waste dumped in the world oceans and other high activity disposals.

Studies by the International Arctic Sea Assessment Project (Sjoeblom and Linsley 1993) concluded that biological radiation doses from anthropogenic radionuclides in the marine environment are generally one to two orders of magnitude less than the doses from such radionuclides in the terrestrial environment (Aarkrog 1998, 2003). A substantial part of the radioactive contamination occurs in the marine environment where by the year 2000 total inventories of Strontium, Cesium and Plutonium in the Oceans were 0.2EBq ^{90}Sr (1 Exabecquerel $=10^{18}$ Becquerel), 0.35 EBq ^{137}Cs and 0.01 Ebq $^{239+240}Pu$. For ^{137}Cs, global fallout contributes 69%, local fallout 21%, reprocessing 7%, and Chernobyl 3% of the total inventory (Aarkrog 2003). The most contaminated ocean region is the Northeast Atlantic Ocean.

Anthropogenic radionuclide concentrations determined in seawater and sediments collected in 1991, 1994 and 1996 in the Eurasian Arctic shelf and interior, are derived from global fallout, from European reprocessing plants and from the Chernobyl accident.[69] The total input of ^{134}Cs, ^{137}Cs and ^{90}Sr from these sources has

[64] https://en.wikipedia.org/wiki/Mayak

[65] http://mentalfloss.com/article/71026/kyshtym-disaster-largest-nuclear-disaster-youve-never-heard

[66] http://www.environmentandsociety.org/arcadia/nuclear-disaster-kyshtym-1957-and-politics-cold-war

[67] http://www.iaea.org/inis/collection/NCLCollectionStore/_Public/29/039/29039179.pdf

[68] http://bellona.org/news/nuclear-issues/radioactive-waste-and-spent-nuclear-fuel/2012-08-russia-announces-enormous-finds-of-radioactive-waste-and-nuclear-reactors-in-arctic-seas

[69] http://www.iaea.org/inis/collection/NCLCollectionStore/_Public/29/039/29039179.pdf

been decreasing during the 1990's, while ^{129}I increased. Radionuclides concentrations decrease with depth. Chernobyl derived ^{137}Cs appeared in the central Arctic Ocean around 1991. The transfer times of radionuclides derived from Sellafield are estimated to be 5–7 years to the SE Barents Sea, 7–9 years to the Kara Sea, 10–11 years to the Laptev Sea and 12–14 years to the central Arctic Ocean. Global fallout is the primary source of plutonium with highest concentrations found in the Atlantic layer of the Arctic Ocean. When transported over the shallow shelf seas, particle reactive transuranic elements experience an intense scavenging. Approximately 75% of the plutonium entering the Kara and Laptev Seas is removed by sediment. High seasonal riverine input of $^{239+240}Pu$ is observed near the mouths of the large Russian rivers. Sediment inventories show much higher concentrations on the shelf compared to the deep Arctic Ocean, due to the low particle flux in the open ocean. Ice cores from glaciers from the Belukha glacier in the Siberian Altai, proximal to the Soviet Arctic nuclear tests, corresponding to the period 1941–1986, display peak Pu concentrations for 1963 consistent with peak ^{3}H activity, coinciding with the Soviet nuclear weapons tests. The shapes of the ^{239}Pu and ^{3}H profiles reflect two main periods of atmospheric nuclear test activity. The former tests contributed about 45% of the integrated Pu inventory. The average $^{240}Pu/^{239}Pu$ ratio of 0.18 ± 0.05 indicates the bulk of the Pu in the Belukha glacier originates from global stratospheric fallout rather than from direct tropospheric input (Olivier et al. 2004).

2.2.2 The US Hydrogen Bomb Tests in the Marshall Islands[70]

During 1946–1958 the US conducted a series of 23 nuclear tests in the Bikini and Eniwetok Atolls at seven test sites on the reef, at sea and in the atmosphere, releasing a combined fission yield of 42.2 Megatons of explosive energy (Fig. 2.10). The series in 1954 set off a 15 megaton thermonuclear hydrogen bomb, about 1000 times the energy of the Hiroshima and Nagasaki atomic bombs. The nuclear fallout contaminated the soil and water of the Bikini atoll (Simon et al. 2010; The Guardian 2014) rendering it uninhabitable. Adult internal thyroid doses for the Utrik community and for the Rongelap Island community were about 760 mSv and 7600 mSv respectively. Compare with 2.42 mSv/year of human exposure to natural background radiation. About 170 excess cancers (radiation-related cases) are projected to occur among more than 25,000 Marshallese, half of whom were born before 1948.

Global stratospheric fallout peaked in 1962 in the Marshall Islands as a delayed consequence of the 1950s tests, while some of the islands remain unlivable.[71] The 15-megatonne Bravo test on 1 March 1954 was a thousand times more powerful than the atomic bomb dropped on Hiroshima, exposing thousands in the surrounding area to radioactive fallout. The high 0.24 ratio of $^{240}Pu/^{239}Pu$ exceeds that of the global

[70] https://www.ctbto.org/nuclear-testing/the-effects-of-nuclear-testing/the-united-states-nuclear-testing-programme/

[71] https://www.theguardian.com/world/2014/mar/02/bikini-atoll-nuclear-test-60-years

Fig. 2.10 During 1946–1958, the Bikini Atoll, part of the Marshall Islands, constituted the site of 23 nuclear detonations, including a Hydrogen bomb test, in 1954 (https://en. wikipedia.org/wiki/Operation_Crossroads#/media/File:Operation_Crossroads_Baker_Edit.jpg)

fallout ratio ($^{240}Pu/^{239}Pu$ = 0.18–0.19). Seawater and sediment samples from the North Pacific exhibit a wide range of $^{240}Pu/^{239}Pu$ values (0.19–0.34), with a trend towards higher ratios in the subsurface waters and sediment. Deep water $^{240}Pu/^{239}Pu$ ratios are higher in the vicinity of the Marshall Islands relative to stations further away. These preliminary data suggest that fallout Pu from the Marshall Islands is more rapidly removed from the surface waters than global Pu fallout (Buesseler 1997).

The consequences of the Marshall Island tests (Fig. 2.11) are detected in the South China Sea (Wu et al. 2014) and Sulu Sea (Dong et al. 2010). The $^{240}Pu/^{239}Pu$ ratios in the sediments of both areas (South China Sea – mean: 0.251; Sulu Sea (Zheng et al. 2012, 2013; Dong et al. 2010) mean 0.280) are higher than the global fallout value of 0.178 ± 0.019, suggesting derivation of Pu from North Pacific nuclear tests. Low inventories of $^{239+240}Pu$ in sediments observed in the South China Sea (3.75 Bq/m2) and the Sulu Sea (1.38 Bq/m2) are still present in the water column. Scavenging and benthic mixing processes were considered to be the main processes controlling the distribution of Pu in the deep-sea sediments of both study areas. (Figs. 2.12 and 2.13).

Recent radiation measurements in the Marshall Islands (Bordner 2016) indicate radiation levels remain high in the Bikini Atoll (Sumner 2016) up to ~800 mRem/year [<8 mSv/year) and to a lesser extent the Nam Atoll and Rongelap Atoll (<100 mRem/year) (<1 mSv/year) (Fig. 2.16) about half a century following the hydrogen bomb tests, which allows estimates of the long term effects of nuclear explosions on the scale of <500 megaton. (Tables 2.4 and 2.5).

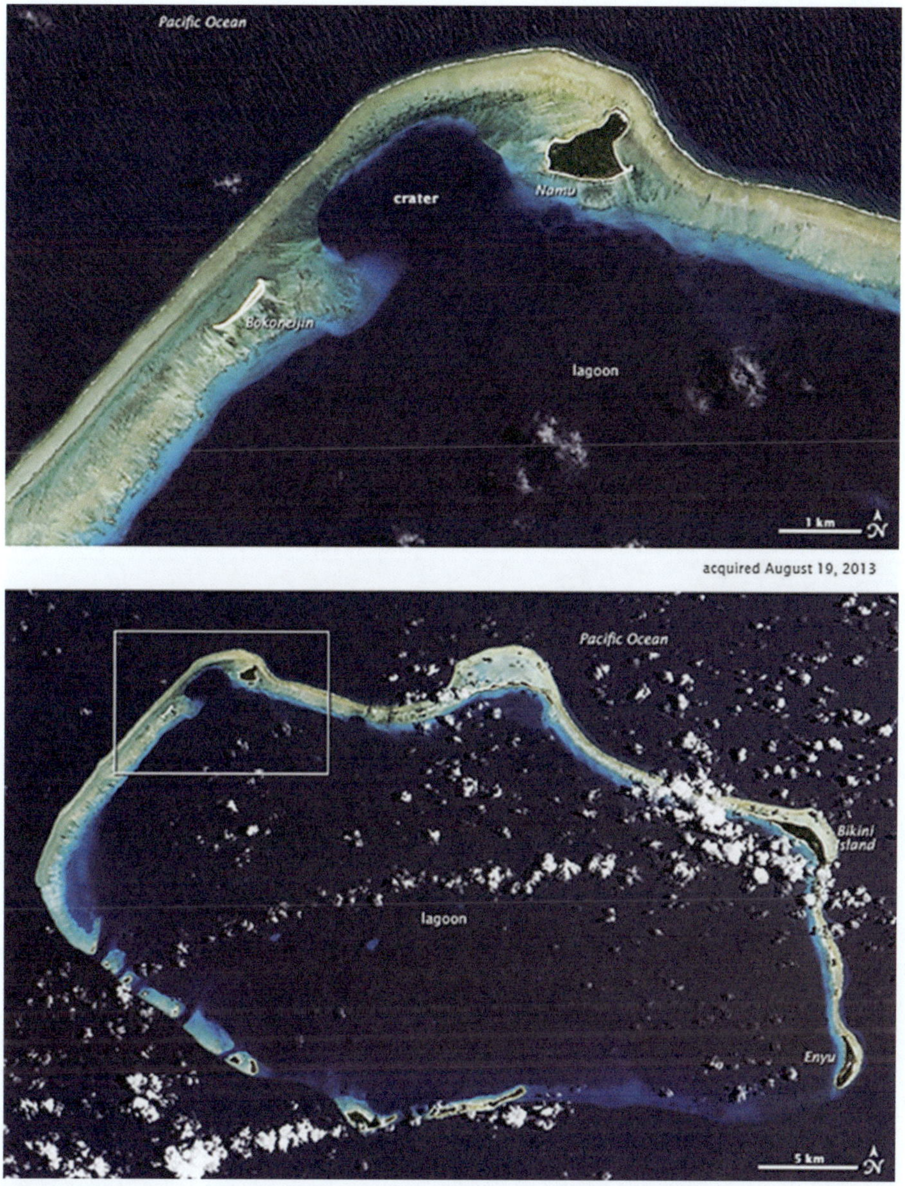

Fig. 2.11 Aerial photographs of the nuclear crater at Namu, Bikini Atoll (https://earthobservatory.nasa.gov/IOTD/view.php?id=83237)

Fig. 2.12 Redwing hydrogen bomb test (Public Domain). *Inset*: Edward Teller "*Father of the Hydrogen bomb*" (https://commons.wikimedia.org/wiki/File:Atombombentest_Redwing-Seminole_02.jpg, https://www.flickr.com/photos/ucdaviscoe/7116427257; https://en.wikipedia.org/wiki/Operation_Redwing)

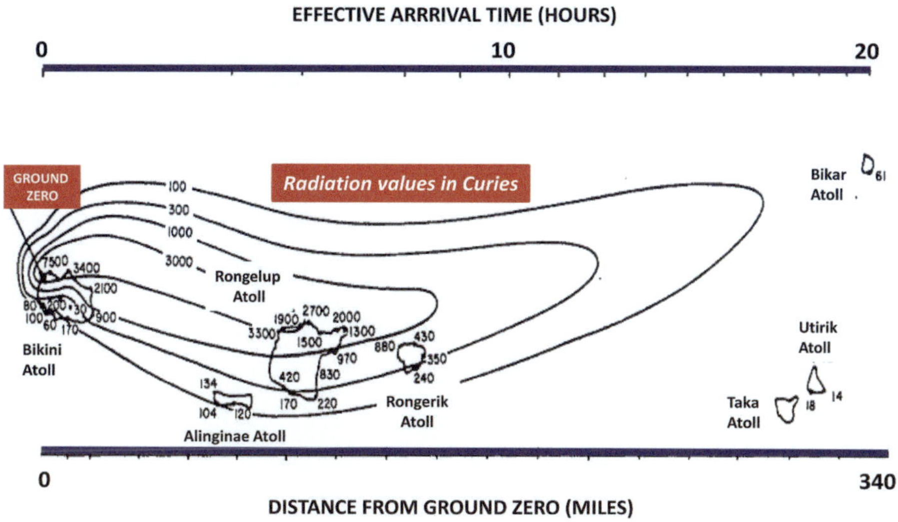

Fig. 2.13 Downwind distribution of fallout from the Castle Bravo Hydrogen bomb test in the Bikini atoll. Values in Curies (https://en.wikipedia.org/wiki/Castle_Bravo#Fallout)

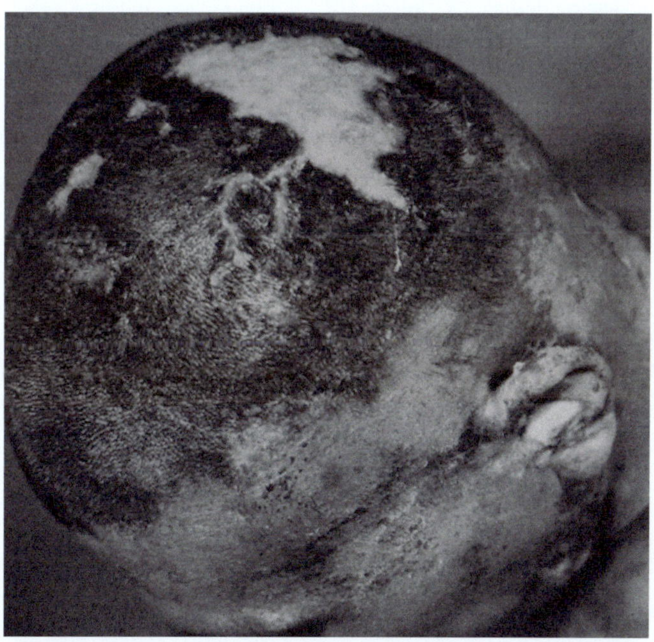

Fig. 2.14 The head of one of the crew members of Daigo Fukuryū Maru, showing radiation burns caused by fallout that collected in his hair. Dated on 7 April 1954, 38 days after the nuclear test (https://en.wikipedia.org/wiki/Nuclear_testing_at_Bikini_Atoll#/media/File:The_head_of_the_crew_of_Daigo_Fukuryu-maru.JPG)

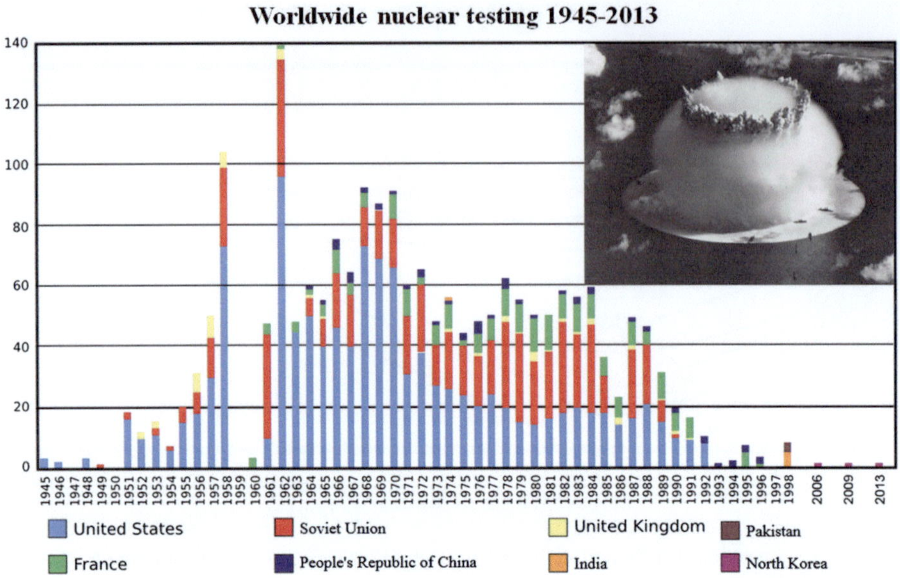

Fig. 2.15 Worldwide nuclear explosions – totals of all nuclear testing 1945–1998 except for the atomic bombings of Hiroshima and Nagasaki. (https://commons.wikimedia.org/wiki/File:Worldwide_nuclear_testing.svg); *Inset*: The "Wilson cloud" from test Baker, situated just offshore from Bikini Island (https://en.wikipedia.org/wiki/Nuclear_testing_at_Bikini_Atoll#/media/File:Crossroads_baker_explosion.jpg)

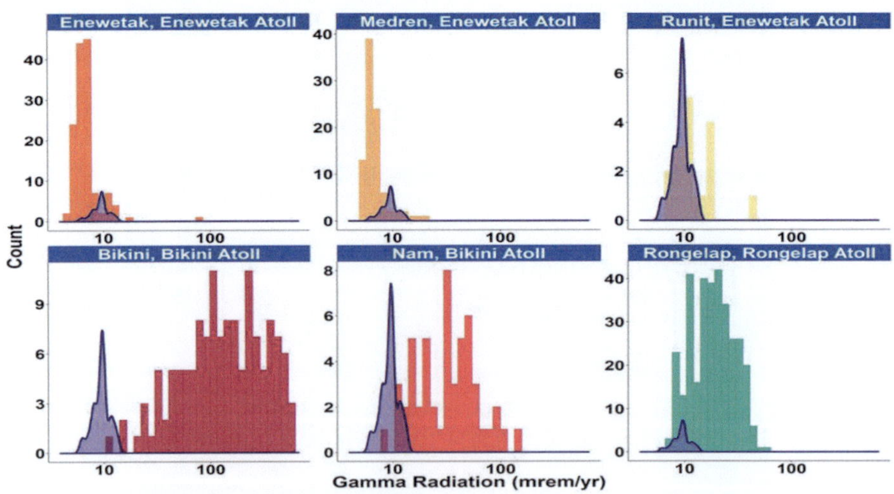

Fig. 2.16 Measurement of background gamma radiation in Millirem; 1 mRem =0.01 mSv) in the northern Marshall Islands (Bordner 2016). The *purple curve* represents the fitted distribution of measured radiation levels on the control island, Majuro. The *vertical axis* is scaled differently for different islands to account for varying radiation levels. Enewetak Island, n = 137; Medren Island, n = 91; Runit Island, n = 20; Bikini Island, n = 137; Nam Island, n = 52; Rongelap Island, n = 332

Table 2.4 Calculated radiation doses at two locations in Rongelap Atoll from fallout following the March 1, 1954 test at Bikini Atoll

	Accumulated dose mSv	
Exposure period after explosion	Inhabited location	Uninhabited location
First 96 h	2200	33,000
96 h to 1 week	350	5300
1 week to 1 month	750	10,800
1 month to 1 year	750	11,000
total to 1 year	4050	6010

1 rad = 10 millisievert

Table 2.5 Major radioactive substances released by the Chernobyl accident with quantities expressed in exabecquerel (*Ebq* = 10^{18} Becquerel). Radioactive isotopes released (http://www.greenfacts.org/en/chernobyl/chernobyl-greenfacts-level2.pdf)

Isotope	Half-life	*EBq*[a]
iodine-131	8.04 days	1.760
cesium-137	30 years	0.085
strontium-90	29.12 years	0.010
Plutonium-241decaying into Americium-241	14.4, 430 years	0.003
Plutonium-238	88 years	
Plutonium-239	24,100 years	
Others		12.140
Total radioactivity released		14

[a]Ebq = 10^{18} Bq

2.2.3 *French Nuclear Tests in Algeria and Polynesia*

Between 1960 and 1965 France conducted 17 nuclear weapons tests, 4 atmospheric and 13 underground, at two locations in Algeria, including a failed test in 1962 at Taourirt Tan Afella where between 5% and 10% of the radioactive material escaped as melt, aerosols and gaseous products (CTBTO[72]). From 1966 tests proceeded in the Moruroa (Fig. 2.17) and Fangataufa atolls in the South Pacific. The atoll of Tureia with around 60 inhabitants, about 100 km away from Moruroa, remained within the zone designated as dangerous. Over the next 30 years a total of 193 atmospheric and underground tests were conducted in the South Pacific. The amount of plutonium-239 dispersed by 45 announced French atmospheric tests is estimated as 6750 curies. Accordingly the amount of dispersed cesium-127 and strontium-90 would have been 1.7 million curies and 1.1 million curies, respectively, with much of the cesium and strontium remaining in the ground and in the water. Underground

[72] https://www.ctbto.org/nuclear-testing/the-effects-of-nuclear-testing/frances-nuclear-testing-programme/

Fig. 2.17 Moruroa Atoll test (https://www.flickr.com/photos/ctbto/9040657030, For a photomontage of nuclear explosions see http://www.theatlantic.com/photo/2015/07/70-years-since-trinity-when-we-tested-nuclear-bombs/398735/)

tests resulted in fracturing of coral, landslides, subsidence, tsunamis and earthquakes. Long-term effects include leakage of fission products to the biosphere and transfer of dissolved plutonium from the lagoon to the ocean and the food chain. Unusual concentrations of iodine-131 were found in marine organisms and of Krypton-85 and tritium in the air and water. These tests have turned the Moruroa atoll into a long-term waste dump, with sediments in the lagoon containing some 20 kg of plutonium and air concentrations of Plutonium-239 about four times greater than in continental France. About 75% of the cesium and strontium remain underground and is leaking to the surface and reaching the food chain, inducing food poisoning (Ruff 2015[73]).

2.2.4 *China's Lop Nur Nuclear Tests*

Commencing on the 16 October 1964, China exploded 45 nuclear bombs including 23 atmospheric and 22 underground explosions of about 1 kilotons to about 4 Megatons at the Lop Nur site, in the Xinjiang region in western China, a region spanning 1.6 million square kilometers and the home of some 20 million people including Uyghur, Han, Kazakh, Hui, Kyrgyz, and Mongol (Fig. 2.18) (CTBTO[74]). The effects on human health, animals and the environment are largely unexplored.

[73] file:///C:/Users/Andrew/Downloads/irc97_15%20(9).pdf

[74] https://www.ctbto.org/nuclear-testing/the-effects-of-nuclear-testing/chinas-nuclear-testing-programme/

Fig. 2.18 A space image of the Lop Nur region in western China (NASA), (https://en.wikipedia.org/wiki/Lop_Nur#/media/File:Basin_of_Lop_Nur_90.25E,_40.10N,_Desert_of_Lop,_Kum_Tagh_and_Astin_Tagh.jpg)

According to the International Physicians for the Prevention of Nuclear War (IPPNW) report Radioactive Heaven and Earth[75] the total amount of released plutonium-239 by 23 atmospheric tests weighing about 48 kg released about 3300 curies. 25 kg Plutonium-239 remaining within the Lop Nur site from underground testing was estimated as 1800 curies. A further two million curies of cesium-137 and 1.3 million curies of strontium-90 were released into the atmosphere. One millionth of a gram of plutonium-239 can cause cancer if inhaled.

According to Dr. Tohti, a Chinese medical doctor, the 1964–1996 nuclear tests at the Lop Nur site in the Xinjian region may have resulted in the deaths of a few hundred thousand people as a consequence of radiation produced from at least 40 nuclear explosions.[76] According to Dr. Takada 46 nuclear tests with a cumulative yield of over 200 megatons affected cities on the ancient Silk Road trade route downwind from Lop Nur, becoming a major contributor to an increase in the incidence of cancer and birth defects (Epoch Times).[77] Meralli (2009), Scientific American 2009 in an article titled '*Did China's Nuclear Tests Kill Thousands and Doom Future Generations? Radioactive clouds hung over villagers as China*

[75] http://ieer.org/wp/wp-content/uploads/1991/06/RadioactiveHeavenEarth1991.pdf

[76] https://www.scientificamerican.com/article/did-chinas-nuclear-tests/

[77] http://www.theepochtimes.com/n3/1524571-chinese-nuclear-tests-allegedly-cause-750000-deaths/

detonated nuclear bombs in the air for four decades'[78] wrote: "*In the early 1990s Jun Takada (a Japanese physicist) who studied radiation effects from tests conducted by the U.S., the former Soviet Union and France, was invited by scientists in Kazakhstan, which borders Xinjiang, to evaluate the hazard from Chinese tests. He devised a computer model to estimate fallout patterns using Soviet records of detonation size and wind velocity as well as radiation levels measured in Kazakhstan from 1995 to 2002. Takada was not allowed into China, so he extrapolated his model and used information about the population density in Xinjiang to estimate that 194,000 people would have died as a result of acute radiation exposure. Around 1.2 million received doses high enough to induce leukemia, solid cancers and fetal damage.*" "*My estimate is a conservative minimum,*" Takada says.

The radio-activation of large areas of Earth does not require deliberate acts of war or tests and has repeatedly occurred by accident, as in the following instances[79]:

13.2.1950: A Convair B-36B crashed in northern British Columbia after jettisoning a Mark IV atomic bomb.

22.5.1957: A 42,000-pound Mark-17 hydrogen bomb accidentally fell from a bomber near Albuquerque, New Mexico.

7.6.1960: The Fort Dix IM-99 accident destroyed a Boeing CIM-10 Bomarc nuclear missile and shelter and contaminated the BOMARC Missile.

24.1.1961: A B-52 Stratofortress carrying two Mark 39 nuclear bombs broke up in mid-air, dropping its nuclear payload in the process.

1965 A-4 crash Philippine Sea. A Skyhawk attack aircraft with a nuclear weapon fell into the sea. The pilot, the aircraft, and the B43 nuclear bomb were never recovered.

17.1.1966: A Palomares crash of a B-52G bomber of the USAF which collided with a KC-135 tanker during mid-air refueling off the coast of Spain. Of the four Mk28 type hydrogen bombs the B-52G carried three were found on land near Almería, Spain.

21.1.1968: B-52 crash involved a United States Air Force (USAF) B-52 bomber carrying four hydrogen bombs when a cabin fire forced the crew to abandon the aircraft. The bomber crashed onto sea ice in Greenland, which resulted in widespread radioactive contamination.

18.9.1980: A Titan missile equipped with a nuclear warhead exploded at Damascus, Arkansas.

Reviews of the modern state of nuclear weapons (Thakur and Evans 2013,[80] Gareth Evans et al. 2015[81]) summarize the state of play as follows: "*By the end of 2009 hopes were higher than for many years that the world was at last seriously headed towards nuclear disarmament. Maybe, just maybe, we could not only stop further proliferation of the most indiscriminately inhumane weapons ever invented,*

[78] https://www.scientificamerican.com/article/did-chinas-nuclear-tests/

[79] https://en.wikipedia.org/wiki/Nuclear_weapon

[80] https://cnnd.anu.edu.au/files/2013/state-of-play-report/Nuclear-Weapons-The-State-of-Play.pdf

[81] https://cnnd.crawford.anu.edu.au/publication/cnnd/5328/nuclear-weapons-state-play-2015

but actually, over time, eliminate them from the face of the globe once and for all. President Barack Obama's Prague Speech of 2009 had set the tone, with its superbly articulated vision of a nuclear-weapon-free world. The report of the International Commission on Nuclear Non-proliferation and Disarmament (ICNND) in 2009, following others before it, had set an achievable global agenda, describing in detail all the building blocks that had to be constructed along the way. In 2009 the United States and Russia were back negotiating nuclear arms control more seriously than they had been for a decade. A major Nuclear Security Summit was planned for 2010, with a sharply practical agenda designed to inhibit both proliferation and nuclear terrorism. And there was every sign, in the lead-up to the 2010 Non-Proliferation Treaty Review Conference, that – utterly unlike its failed predecessor five years earlier – there would be consensus for significant forward movement across the whole spectrum of inter-related disarmament, non-proliferation and peaceful-uses issues. By the end of 2012, however, much of the sense of optimism of three years earlier had evaporated. Certainly some progress had been made, and on a few issues, on the face of it, quite substantial progress. The New START treaty, signed by the United States and Russia in 2010, will significantly reduce the number of deployed strategic weapons. The 2010 US Nuclear Posture Review did make some moves in the direction of reducing reliance on nuclear weapons. The 2010 NPT Review Conference succeeded in reaching agreement on 64 action points (a refreshing change from zero in 2005), adopted strong new language on the catastrophic humanitarian consequences of the use of nuclear weapons, and supported initial moves towards a weapons-of-mass-destruction-free zone in the Middle East. And at the Nuclear Security Summits (NSS) in both 2010 and 2012, states made strong commitments to ensure that weapon-useable materials, and weapons themselves, do not fall into the hands of rogue states or terrorists. But New START left both US and Russian stockpiles intact, their high-alert status undisturbed, weapons-modernization programs in place, disagreements about missile defense and conventional arms imbalances unresolved – and talks on further drawdowns going nowhere. Nuclear weapons numbers have decreased overall, as a result of actions by the United States and Russia in particular, but there has been an actual acceleration of nuclear-weapons programs in India, Pakistan, and China. The cautious initial doctrinal move by Washington towards accepting that the "sole purpose" of nuclear weapons is to respond to nuclear threats, not those of any other kind, has gone nowhere, inhibited by resistance from its Northeast Asian and more nervous Central and Eastern European allies."

By the first part of the twenty-first century rising tensions between global powers and increasing risk of nuclear war, including accidental nuclear war (Phillips 1998/2017)[82,83,84]

[82] http://nuclearfiles.org/menu/key-issues/nuclear-weapons/issues/accidents/20-mishaps-maybe-caused-nuclear-war.htm

[83] http://futureoflife.org/background/nuclear-close-calls-a-timeline/

[84] https://en.wikipedia.org/wiki/List_of_nuclear_close_calls

have left little room for optimism,[85,86,87,88] with a consequent shift of the clock of the atomic scientists on the 27 January, 2017, to 2.5 min to midnight.[89]

2.5 min to midnight, from the Bulletin of the Atomic Scientists (http://thebulletin.org/)

2.3 Late Anthropocene Radioactive Mausoleums

The radioactive contamination of Earth does not need nuclear wars—it has already commenced with nuclear tests, leaks and accidents in power plants, and nuclear dumping on land and at sea, leaving live radioactive mausoleums destined to last for some 24,100 years, the half-life of Plutonium, such as at Chernobyl and Fukushima (Figs. 2.19 and 2.20):

2.3.1 Chernobyl

The Radioactive Cross

Millennia ago
I strung the via Apia
Hugging dying slaves
Of Spartacus uprising
Crucified by Roman legionnaires
With the brutality of
Empire

[85] http://www.telesurtv.net/english/news/US-Has-Enough-Nuclear-Weapons-to-Destroy-Itself-4-Times-Over---20140914-0033.html

[86] http://fpif.org/may-greater-risk-nuclear-catastrophe-cold-war/

[87] http://www.cbsnews.com/news/60-minutes-risk-of-nuclear-attack-rises/

[88] http://www.ucsusa.org/nuclear-weapons/us-china-relations/risk-nuclear-war-china#.WIP2-FN96Uk

[89] http://thebulletin.org/timeline

Fig. 2.19 Radiation hot spots over Ukraine and Belarus in the wake of the Chernobyl nuclear accident (137-Cs in Curies/km2). (1 curie = 37.109 Becquerel) (https://upload. wikimedia.org/wikipedia/commons/2/23/Chernobyl_radiation_map_1996.svg)

Two thousand years before
I bore your saviour
A carpenter from Nazareth
Crucified to save you
From yourselves and offer
God's love to this tormented
Land

Centuries since
I beamed on you from every wall
In promise of salvation
From life toll's miseries
Enshrined in the hereafter

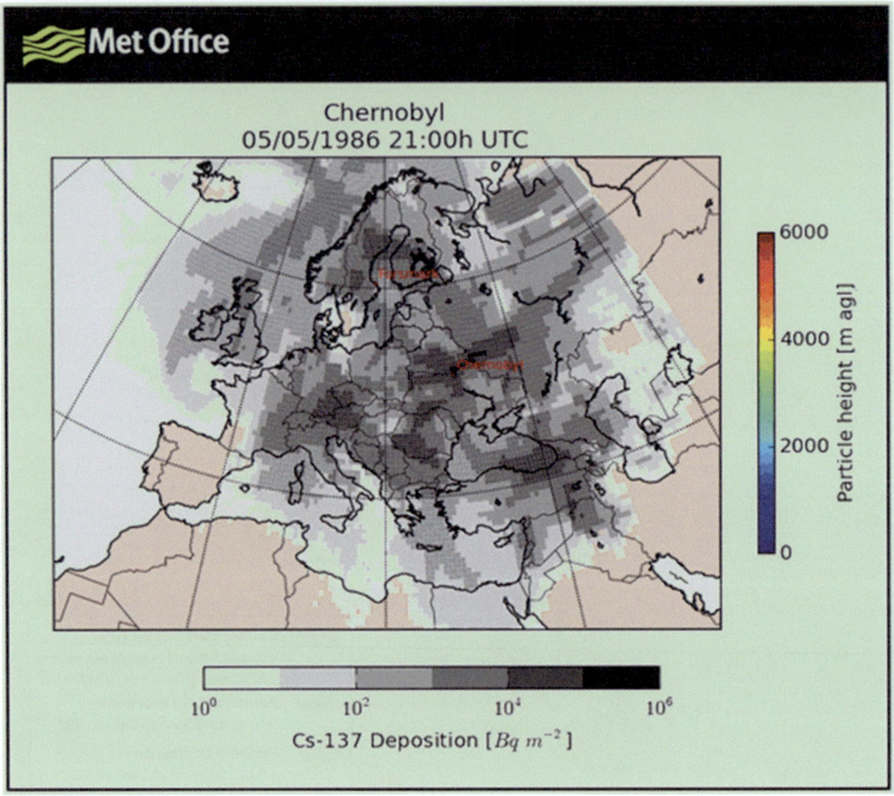

Fig. 2.20 Cs-137 Deposition (Bq m−2) in Europe following the Chernobyl nuclear accident (https://metofficenews.files.wordpress.com/2011/04/chernobylcaesium-600.gif). With permission of the UK Meteorological Office 2011

Pointing a road to
Heaven

Years hence
You waved me on your spider flags
In bloody crusade battlefields
Presiding over rape and murder
Turning my light to dark
My love into hate flames of
Hell

Nowadays
I shine again
With deadly gamma rays
Destroying my believers
Atom-smashing Prometheans
A torch of warning for all time in
Empty space.

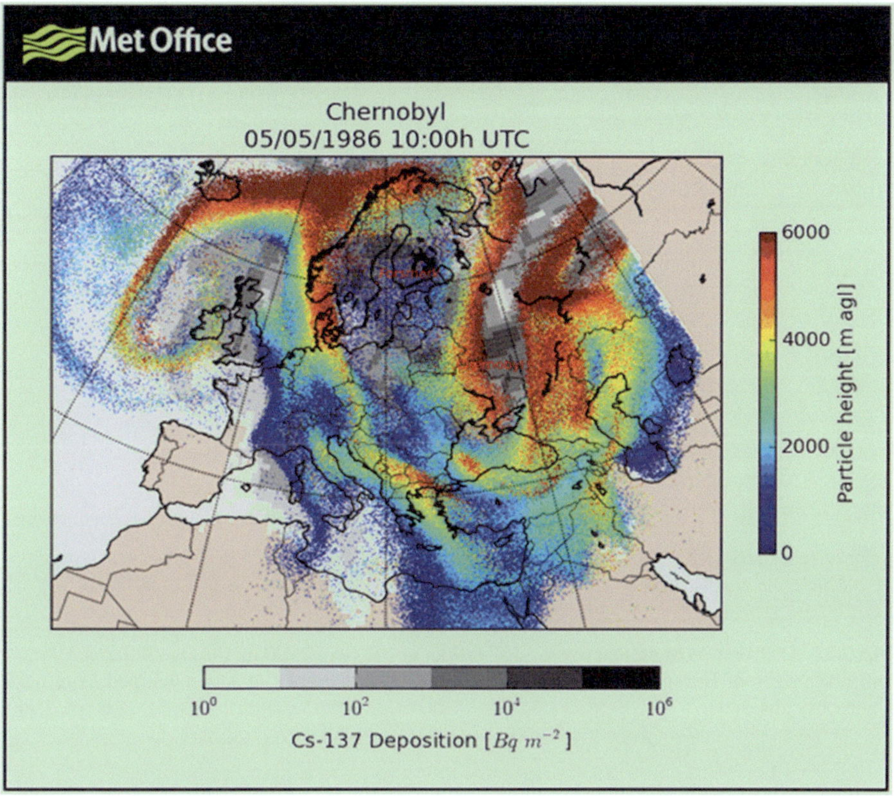

Fig. 2.21 Radioactive particle height above ground level in the wake of the Chernobyl nuclear accident (https://metofficenews.files.wordpress.com/2011/04/chernobylcaesium-600.gif). With permission of the UK Meteorological Office 2011

The catastrophic nuclear accident in the Chernobyl Nuclear Power Plant, Pripyat, Ukraine, on 26 April 1986, involved a steam explosion and fire of the exposed graphite moderator, releasing radioactive particles over much of the western USSR and Europe (Figs. 2.21 and 2.22) (Green Facts),[90] an event classified as a level 7 event—the highest classification on the International Nuclear Event Scale.[91] Major releases of radionuclides reached 7.3 million curies,[92] continuing for 10 days, and including radioactive gases, condensed aerosols and a large amount of fuel particles. The total release of radioactive substances was about 14 Ebq [1 EBq = 10^{18} Becquerel]. The noble gases contributed about 50% of the total release. More than 200,000 Km^2 of Europe received levels of Cs-137 above 37,000 Bq/m^2. Over 70% of this area was in the three most affected countries, Belarus, Russia and Ukraine

[90] http://www.greenfacts.org/en/chernobyl/l-3/3-chernobyl-environment.htm

[91] http://www-ns.iaea.org/tech-areas/emergency/ines.asp

[92] http://www.tinyvital.com/Misc/nukes.htm

Areas of Europe contaminated with ^{137}Cs

Country	37–185 k Bq/m^2		185–555 kBq/m^2		555–1480 kBq/m^2		>1480 kBq/m^2	
	km^2	% of country	km^2	% of country	km^2	% of country	km^2	% of country
Belarus	29,900	14.4	10,200	4.9	4,200	2.0	2,200	1.1
Ukraine	37,200	6.2	3,200	0.53	900	0.15	600	0.1
Russia	49,800	0.29	5,700	0.03	2,100	0.01	300	0.002
Sweden	12,000	2.7	–	–	–	–	–	–
Finland	11,500	3.4	–	–	–	–	–	–
Austria	8,600	10.3	–	–	–	–	–	–
Norway	5,200	1.3	–	–	–	–	–	–
Bulgaria	4,800	4.3	–	–	–	–	–	–
Switzerland	1,300	3.1	–	–	–	–	–	–
Greece	1,200	0.91	–	–	–	–	–	–
Slovenia	300	1.5	–	–	–	–	–	–
Italy	300	0.1	–	–	–	–	–	–
Moldova	60	0.2	–	–	–	–	–	–
Totals	**162,160 km^2**		**19,100 km^2**		**7,200 km^2**		**3,100 km^2**	

Fig. 2.22 Distribution of radioactive Cs-137 fallout in 103 Bequerel/km^2 through Belarus, Ukraine and other parts of Europe in the wake of the Chernobyl disaster (https://en.wikipedia.org/wiki/Chernobyl_disaster)

(Figs. 2.21 and 2.22). The deposition was extremely varied and was enhanced in areas where it was raining. Most of the strontium and plutonium radioisotopes were deposited within 100 km of the destroyed reactor due to larger particle sizes.

According to the Chernobyl Gallery typical dosage recorded in those Chernobyl workers who died within a month was 6000 mSv, Chernobyl workers received an average dose of 170 mSv/h and exposure of Chernobyl residents who were relocated after the blast in 1986 was 350 mSv. Of all the children exposed by drinking milk from 1986 to 2002 about 4000 were diagnosed with thyroid cancer. As of September 2005, 15 had died.

Although the effects of the Chernobyl disaster on humans have been comprehensively documented a controversy exists regarding the long term health effects on the human population (Nuclear Energy Association,[93] World Nuclear Association

[93] http://www.world-nuclear.org/information-library/safety-and-security/safety-of-plants/chernobyl-accident.aspx

(2012, 2017),[94] Hatch et al. (2005), Leitner (2004), NEI (Nuclear Energy Institute 2015)[95]). The main consequences according to the World Health Organization[96]:

1. 28 fire fighters and staff and the reactor were killed and more than 560 heavily exposed to radiation.
2. More than 400,000 clean-up workers were exposed to radiation.
3. More than 270,000 living in areas of Belarus, Russia and Ukraine were contaminated by radioactivity.
4. More than 116,000 people were evacuated from heavily contaminated to safe areas.
5. Many children were exposed to radio-iodine (more than 2000 thyroid cancers in children)

The Chernobyl Forum (2003–2005)[97] estimates the average doses of radiation doses for emergency workers (liquidators) (1986–1989), evacuees from highly contaminated zones (1986), residents of strict-control areas (1986–2005) and residents from other contaminated areas well above the permissible upper limits for nuclear workers. The report[98] states: *The projections indicate that, among the most exposed populations (liquidators, evacuees and residents of the so-called 'strict control zones'), total cancer mortality might increase by up to a few per cent owing to Chernobyl related radiation exposure. Such an increase could mean eventually up to several thousand fatal cancers in addition to perhaps one hundred thousand cancer deaths expected in these populations from all other causes.*

Radioactive contamination from the Chernobyl meltdown spread over 40% of Europe, including Austria, Finland, Sweden, Norway, Switzerland, Romania, Great Britain, Germany, Italy, France, Greece, Iceland, Slovenia, and wide territories in Asia including Turkey, Georgia, Armenia, Emirates, China, northern Africa, and North America.[99] Nearly 400 million people resided in territories that were contaminated with radioactivity at a level higher than 4000 Bq/m^2 (0.11 Ci/km^2) from April to July 1986. Nearly 5 million people, including more than 1 million children, still live with dangerous levels of radioactive contamination in Belarus, Ukraine and European Russia.[100] In the wake of the Chernobyl reactor catastrophe radioactive fallout has risen from a background of approximately 1000 Bq/m2 of cesium-137 per square meter over 100,000 square km through Europe to radioactivity values of >1480.10^3 Bq/m2 over 3100 km^2, 555–148.10^3 Bq/m2 over ~7200 km^2, and 185–555.10^3 Bq/m2 over 19,100 km^2 over Belarus, Ukraine and Russia. Further spread

94 http://www.world-nuclear.org/information-library/safety-and-security/safety-of-plants/appendices/chernobyl-accident-appendix-2-health-impacts.aspx

95 https://www.nei.org/master-document-folder/backgrounders/fact-sheets/chernobyl-accident-and-its-consequences

96 http://www.who.int/ionizing_radiation/chernobyl/Overview_WHO_past_involvement.pdf

97 https://www.iaea.org/sites/default/files/chernobyl.pdf

98 https://www.iaea.org/sites/default/files/chernobyl.pdf

99 https://en.wikipedia.org/wiki/Chernobyl_disaster

100 http://www.ncbi.nlm.nih.gov/pubmed/20002040

of Cesium contamination into western and southern Europe led to rises of ~37–185.10^3 Bq/m^2. In Northern Germany radioactivity rose to approximately 4000–5000 Bq/m^2 while in southern Germany radioactivity rose to around 40,000 Bq/m^2 of radio-cesium. Studies carried out in surrounding countries indicate that over one million people could have been affected by radiation (Yablokov and Nesterenko 2009). Radioactive fallout was enhanced in mountainous regions such as the Alps, the Welsh mountains and the Scottish Highlands, where adiabatic cooling caused radioactive rainfall. Scandinavia has also received heavy fallout when contaminated air collided with a cold front, bringing rain. Europe-wide effects of the Chernobyl are indicated in Fig. 2.21. Note the very high levels of radiation (37–185.10^3 Bq/km^2) over 29,000 km^2 to extreme levels of radiation (>10^3.1480 Bq/km^3) over to 2200 km^2 of Belarus (compare with Tables 2.1 and 2.2).

2.3.2 *Fukushima*

The 2011 Power Plant-1 Fukushima Daiichi nuclear accident (FDNA) in northern Japan's Pacific coast represents the second worst nuclear accident in the history of nuclear power generation. At the time of the accident, only reactors 1–3 were operational, and reactor 4 served as temporary storage for spent fuel rods.[101] The accident released highly volatile fission products including ^{129m}Te, ^{131}I, ^{134}Cs, ^{136}Cs and ^{137}Cs widely distributed in eastern Japan (Fig. 2.23). Information regarding the release of non-volatile actinides, in particular, Pu isotopes, is reported in northwest and south of the Fukushima DNPP in the 20–30 km zones (Zheng et al. 2012). The high activity ratio of ^{241}Pu/$^{239+240}$Pu (>100) from the Fukushima DNPP accident highlights the need for long-term assessment of the ^{241}Pu dose and the growth of ^{241}Am and serious contamination of the environment (Ryall 2011; Guosheng et al. 2015).[102] Measurements of $^{239+240}$Pu activity and ^{240}Pu/^{239}Pu atom ratio in soils from central-eastern Japan for the period 1969–1977 allow definition of a baseline for the pre-Fukushima accident regarding $^{239+240}$Pu of 0.004–1.46 mBq g^{-1}, and ^{240}Pu/^{239}Pu atom ratios of 0.148–0.229 with a mean of 0.186 ± 0.015. For the Fukushima-accident released ^{241}Pu a similar radiation impact is estimated as for the global fallout of ^{241}Pu in the last decades.[103] Estimates of the amount of Pu239 + 240 released from the FDNA accident are given as 1.0–2.4.10^9 Becquerel as compared to Chernobyl where 8.7.10^{13} Bq was released (Zhen et al. 2012) (Fig. 2.24).

The $^{239+240}$Pu activities and ^{240}Pu/^{239}Pu atom ratios in sediments of the northern South China Sea were found to be sourced from both global fallout and current-transported fallout from the Marshall Islands Pacific Proving Grounds (PPG) into the South China Sea through the Luzon Strait,[104] dominated the Pu input in the

[101] https://www.britannica.com/event/Fukushima-accident

[102] https://www.ncbi.nlm.nih.gov/pubmed/23899337

[103] https://www.ncbi.nlm.nih.gov/pmc/articles/PMC4399503/

[104] http://pubs.acs.org/doi/abs/10.1021/es405363q?journalCode=esthag

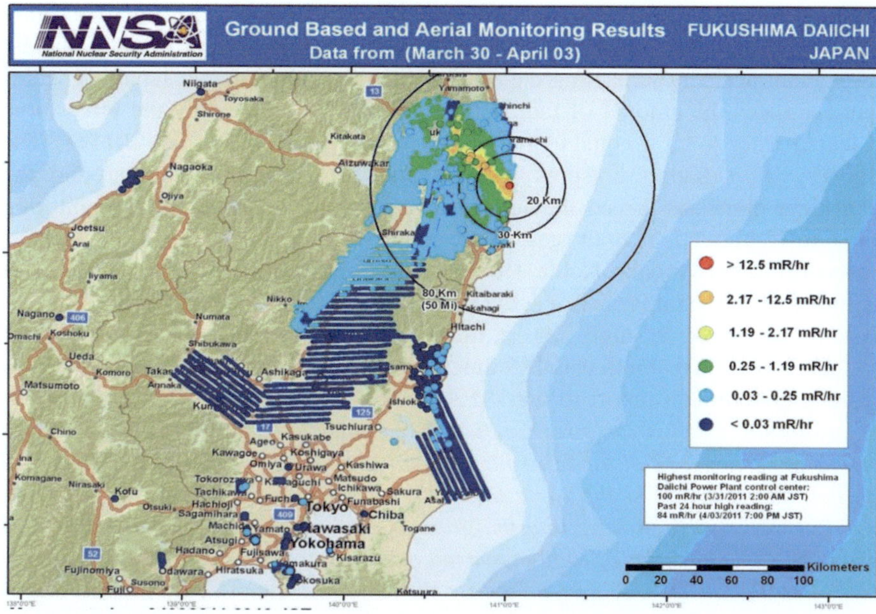

Fig. 2.23 Map of contaminated areas around the plant (22 March–3 April 2011) (https://en.wikipedia.org/wiki/Fukushima_Daiichi_nuclear_disaster#/media/File:NIT_Combined_Flights_Ground_Measurements_30Mar_03Apr2011_results.jpg)

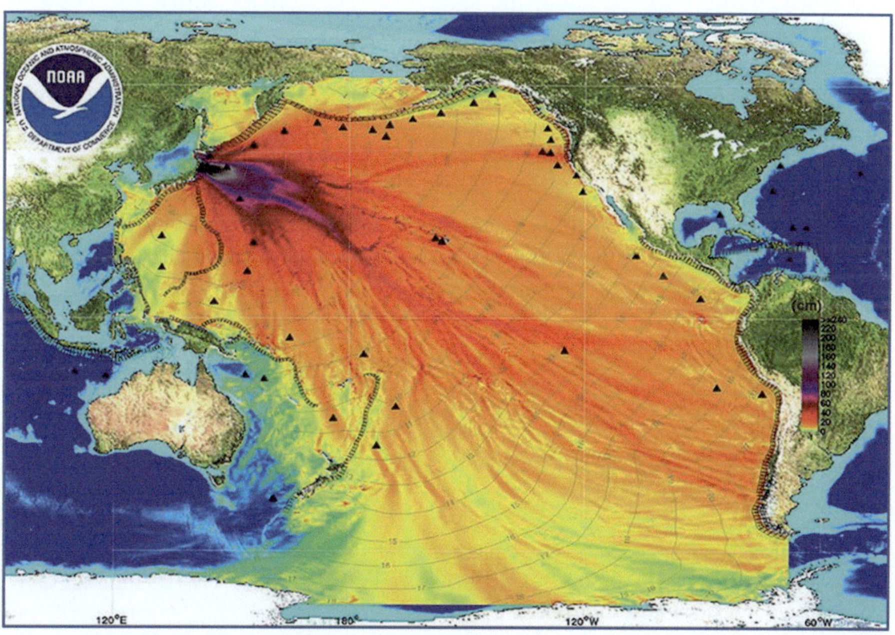

Fig. 2.24 Model simulation of the long term dispersal of Cs-137 into the Pacific Ocean off Fukushima (http://www-mo.tsunami.gov/events/11Mar2011factsheet.php)

1950s indicated by elevated $^{240}Pu/^{239}Pu$ atom ratios (>0.30) in sediments. The study confirmed there were no clear signals of Pu from the Fukushima Daiichi Nuclear Power Plant accident impacting the South China Sea. By contrast Japan's government estimates the amount of radioactive cesium-137 released by the Fukushima nuclear disaster so far is equal to that of 168 Hiroshima bombs (Luisa 2011).[105] The discovery of radiation hotspots well beyond the exclusion zone around the Fukushima plant has forced the Japanese government to increase its monitoring from 6 to 22 prefectures in the east of the country.[106]

According to Hiroaki Koide (Koide 2013), Master of Science in Nuclear Engineering, Assistant Professor at the Kyoto University Research Institute, Nuclear Waste Management & Safety Expert, the cesium-137 released into the atmosphere by Units 1–3 at Fukushima was 168 times that of the Hiroshima bomb, stating in a Japanese government report to the IAEA[107]: "*I myself believe this is probably an underestimate, and two or three times that amount, that is, 400 to 500 times the amount of cesium-137 of the Hiroshima atomic bomb has already been dispersed into the atmosphere.*" *At the same time, the radioactive materials that were dissolved in water were flowing into the ground, then out into the ocean. "I believe almost the same amounts of radioactive material released into the air has probably flowed into the ocean"*. A Tokyo Electric Power Company (TEPCO) report calculates the amount of Cesium leaked from Fukushima is equal to 4023 Hiroshima bombs. TEPCO estimated 15,000 terabecquerels (10^{12} Bq) of cesium radiation have leaked from the Fukushima Nuclear Power Plant in Japan.[108]

The large discharge of radioactivity into the northwest Pacific Ocean has generated considerable concern about the spread of radioactivity across the ocean to North America (Lee 2013). Time series measurements of ^{134}Cs and ^{137}Cs in seawater revealed an initial arrival of the Fukushima signal by ocean current transport at a location 1500 km west of British Columbia, Canada, in June 2012, about 1.3 years after the accident (Smith et al. 2015). By June 2013, the Fukushima signal had spread onto the Canadian continental shelf, and by February 2014, it had increased to a value of 2 Bq/m3 throughout the upper 150 m of the water column, resulting in an overall doubling of the fallout background from atmospheric nuclear weapons tests. Future total levels of ^{137}Cs from Fukushima and nuclear test fallout off the North American coast will likely attain maximum values in the 3–5 Bq/m^3 range by 2015–2016.

[105] http://www.naturalnews.com/033460_Fukushima_Hiroshima.html

[106] http://www.telegraph.co.uk/news/worldnews/asia/japan/8721630/New-radiation-hotspots-prompt-Japan-to-extend-monitoring.html

[107] http://enenews.com/professor-fukushima-release-equivalent-1000-atomic-bombs-video

[108] http://alexanderhiggins.com/fukushima-nuclear-cesium-fallout-now-equals-4023-hiroshima-bombs/

References

Aarkrog A (1998) A retrospect of anthropogenic radioactivity in the global marine environment. Radiat Prot Dosimetry 75(1–4):23–31. Oxford Academic Press, Oxford

Aarkrog A (2003) Input of anthropogenic radionuclides into the World Ocean. Deep-Sea Res II Top Stud Oceanogr 50:2597–2606

Bauer S (2005) Radiation exposure due to local fallout from Soviet atmospheric nuclear weapons testing in Kazakhstan: solid cancer mortality in the Semipalatinsk historical cohort 1960–1999. Radiat Res 164:409–419

Bordner AS (2016) Measurement of background gamma radiation in the northern Marshall Islands. Proc Natl Acad Sci 113(25):6833–6838

Buesseler KO (1997) The isotopic signature of fallout in the North Pacific. J Environ Radioact 36:69–83

Crutzen PJ, Birks JW (1982) The atmosphere after a nuclear war: twilight at noon. Ambio 11(2/3):114–125

CTBTO (Comprehensive Nuclear-Test-Ban Treaty) (2012) https://www.ctbto.org/specials/who-we-are/

Cutter GA et al (1979) Deposition and accumulation of plutonium isotopes in Antarctica. Nature 279:628–629

Dong W, Zheng J, Guo Q, Yamada M, Pan S (2010) Characterization of plutonium in deep-sea sediments of the Sulu and South China Seas. J Environ Radioact 8:622–629

Evans G, Ogilvie-White T, Ramesh T (2015) Nuclear weapons the state of play 2015. Australian National University, Canberra

Global Nuclear Arsenals (2009) Approximately 23,335 weapons with ~ 6400 MT (megatons) yield. http://www.nucleardarkness.org/include/nucleardarkness/files/global_nuclear_arsenal_in_number_2009.pdf

Gould J (1990) Wonderful life: the burgess shale and the nature of history. Angus and Robertson, Sydney, p 347

Guosheng Y et al (2015) Plutonium concentration and isotopic ratio in soil samples from central-eastern Japan collected around the 1970s. Scientific reports US National Library of Medicine, National Institutes of Health. Sci Rep 5:9636

Hancock JJ (2014) The release and persistence of radioactive anthropogenic nuclides. Geol Soc Lond 395:265–281

Hardy EP et al (1973) Global inventory and Distribution of fallout plutonium. Nature 241:444–445

Hatch H et al (2005) The Chernobyl disaster: cancer following the accident at the Chernobyl nuclear plant. Epidemiol Rev 27(1):56–66

Hyperphysics (2016) Department of Physics and Astronomy, Georgia State University. http://hyperphysics.phy-astr.gsu.edu/hbase/NucEne/bomb.html

IEER (Institute for Energy and Environmental Research) (1991) Radioactive heaven and Earth. http://ieer.org/wp/wp-content/uploads/1991/06/RadioactiveHeavenEarth1991.pdf

Ivanov A (1996) 1950's Soviet A-bomb tests still claim victims. Albion Monitor/News. http://www.albionmonitor.com/9611a/russia1950nuke.html

Johnston R (2008) Multi-megaton tests: the largest Nuclear tests. http://www.johnstonsarchive.net/nuclear/tests/multimegtests.html

Josefsson D (1998) Anthropogenic radionuclides in the Arctic Ocean – Distribution and pathways. Lund University. http://www.iaea.org/inis/collection/NCLCollectionStore/_Public/29/039/29039179.pdf

Koide H (2013) Japan Professor: I believe airborne release of cesium-137 from Fukushima equals 400 to 500 Hiroshima nuclear bombs — another 400 to 500 bombs worth has already flowed into Pacific Ocean. ENE news, march 13th, 2013

Lee J (2013) Ten years of Fukushima radiation crossing the Pacific Ocean. CVnews 22 October. https://climateviewer.com/2013/10/22/ten-years-of-fukushima-radiation-crossing-the-pacific-ocean/# prettyPhoto

Leitner K (2004) WHO's involvement in the Chernobyl accident: past and present. http://www.who.int/ionizing_radiation/chernobyl/Overview_WHO_past_involvement.pdf
Lomax ME (2013) Biological consequences of radiation-induced DNA damage: relevance to radiotherapy. Clin Oncol 25(10):578–585
Luisa C (2011) Fukushima radioactive cesium leaks 'equal 168 Hiroshimas. Natural News. http://www.naturalnews.com/033460_Fukushima_Hiroshima.html
Meralli Z (2009) Did China's nuclear tests kill thousands and doom future generations? Sci Am. July 2009. https://www.scientificamerican.com/article/did-chinas-nuclear-tests/
McVicker C (1954) ICE case studies, Novaya Zemlya. http://www1.american.edu/ted/ice/novalya.htm
Miklós V (2013) The tragic story of the Semipalatinsk Nuclear Test Site. http://io9.gizmodo.com/5988266/the-tragic-story-of-the-semipalatinsk-nuclear-test-site
NEI (Nuclear Energy Institute) (2015) Fact sheets. https:// www.nei.org/master-document-folder/backgrounders/fact-sheets/chernobyl-accident-and-its-consequences
Olivier S et al (2004) Plutonium from global fallout recorded in an ice Core from the Belukha glacier, Siberian Altai. Environ Sci Technol 38(24):6507–6512
Pearce F (2015) Shocking state of world's riskiest nuclear waste site. New Scientist. 21 January 2015. https://www.newscientist.com/article/mg22530053-800-shocking-state-of-worldsriskiest-nuclear-waste-site/
Phillips AF (1998/2017) 20 mishaps that might have started accidental nuclear war. Nuclear Files Org. http://nuclearfiles.org/menu/key-issues/nuclear-weapons/issues/accidents/20-mishaps-maybe-caused-nuclear-war.htm
Ploughshares (2016) World Nuclear Weapon Stockpile. http://www.ploughshares.org/world-nuclear-stockpile-report
Rabi T (2012) The nuclear disaster of Kyshtym 1957 and the politics of the cold war. Arcadia – Environment and Society Portal. http://www.environmentandsociety.org/arcadia/nuclear-disaster-kyshtym-1957-and-politics-cold-war
RADNET (2016) (since 1996) Information about source points of anthropogenic radioactivity. An information resource for persons interested in the public safety consequences and radio-ecological impact of nuclear accidents and industries. http://www.davistownmuseum.org/cbm/Rad14.html
Ruff TA (2015) The humanitarian impact and implications of nuclear test explosions in the Pacific region. Int Rev Red Cross 97(899):775–813
Ryall J (2011) New radiation hotspots prompt Japan to extend monitoring. The Telegraph. http://www.telegraph.co.uk/news/worldnews/asia/japan/8721630/New-radiation-hotspots-prompt-Japan-to-extend-monitoring.html
Simone D et al (1979) The effects of nuclear war. Nuclear War Effects Project Library of Congress Catalog Card Number 79–600080. http://atomicarchive.com/Docs/pdfs/7906.pdf
Simon SL et al (2010) Radiation doses and cancer risks in the Marshall islands associated with exposure to radioactive fallout from Bikini and Enewetak nuclear weapons tests: summary. Health Phys 99(2):105–123
Sjoeblom K Linsley GS (1993) International Arctic Seas Assessment Project (IASAP). http://www.iaea.org/inis/collection/NCLCollectionStore/_Public/28/041/28041350.pdf
Smith JN et al (2015) Arrival of the Fukushima radioactivity plume in North American continental waters. Proc Natl Acad Sci 112(5):1310–1315
Sumner T (2016) Bikini Atoll radiation levels remain alarmingly high. Sci News 190(1):16
The Chernobyl Forum (2003–2005) Chernobyl's legacy: health, environmental and socio-economic impacts and recommendations to the Governments of Belarus, the Russian Federation and Ukraine. https://www.iaea.org/sites/default/files/chernobyl.pdf
The Guardian (2014) Marshall Islands Bikini Atoll nuclear test: 60 years later and islands still unliveable. https://www.theguardian.com/world/2014/mar/02/bikini-atoll-nuclear-test-60-years
Turco RP et al (1983) Nuclear winter: global consequences of multiple nuclear explosions. Science 222:1283–1292

UK Meteorological Office (2011) 25 years on from Chernobyl. https://blog.metoffice.gov.uk/2011/04/26/25-years-on-from-chernobyl/

World Nuclear Association (2012) Radiation and life. http://www.world-nuclear.org/information-

World Nuclear Association (2017). http://www.world-nuclear.org/

World Nuclear Weapon Stockpile. http://www.ploughshares.org/world-nuclear-stockpile-report

Wu J et al (2014) Isotopic composition and Distribution of plutonium in Northern South China Sea sediments revealed continuous release and transport of Pu from the Marshall Islands. Environ Sci Technol 48(6):3136–3144

Yablokov AV, Nesterenko VB (2009) Chernobyl contamination through time and space. https://www.ncbi.nlm.nih.gov/pubmed/20002040

Zheng J et al (2012) Isotopic evidence of plutonium release into the environment from the Fukushima DNPP accident. Scientific Reports 2, Article number: 304. http://www.nature.com/articles/srep00304

Zheng J et al (2013) Release of plutonium isotopes into the environment from the Fukushima Daiichi nuclear power plant accident: what is known and what needs to be known. Environ Sci Technol 47(17):9584–9595

Chapter 3
The Event Horizon

Prometheus
Where will you come from rocket
Lands of nuclear crusaders
With God on their side, or
Beleaguered revolutionary raiders
Frozen in steppes, despair in red eyes, or
Napoleonic Zealots, grandeur
Atom merchants masked by lies, or
A teenage soldier Allah's avenger
Heaven's keys on neck promise rebirth, or
From a failed computer chip
A species inheriting Earth

When will you burst warhead
At noon, clouding the blue sky
Squashing a rainbow and verse in my head, or
Late evening storm, a night bird's cry
At dawn, a butterfly's one day life
When two suns will rise, or
On a moon-bewitched night
Stars watching a final demise
Through myriad eyes, tearful lights
Bidding farewell, Adam's children's sojourn
Fading smoke palls cast a veil over morn.

What will you shatter blast wave
In Kali's death dance embrace
A saint's ivy-covered stone grave
Where countless souls sought solace
A green nursery shrouded in fog
A log cabin a forester built in a tree
On a rainy day, for himself and his dog
A sky-scraping Manhattan bank, Mydas glee
Of mammon-worship Beelzebub's gore
The school down the road will no more
Shelter my daughter who entered its door.

The Event Horizon: A metaphor for life in the twenty-first century. In astronomy the Event Horizon constitutes the outer boundary of a black hole in the core of a galaxy. All matter and energy that fall into a black hole through the Event Horizon cannot survive and no radiation, including light, can escape. http://www.britannica.com/topic/event-horizon-black-hole

A.Y. Glikson, *The Plutocene: Blueprints for a Post-Anthropocene Greenhouse Earth*, Modern Approaches in Solid Earth Sciences 13,
DOI 10.1007/978-3-319-57237-6_3

What will you burn fireball?
Wings of an eagle that just tried to soar
An olive tree that's seen empires rise and fall
An indigo sari an Indian girl wore
A broken jar at her feet by the bore
Crying for help, no one heeding her call
Dust-covered tomes of unheeded lore
Mona Lisa's portrait, solemn faces on wall
As if they knew what's in store
Young girl's bronze hair as she falls
Clutching a burning poem on love

Who will you kill neutron ray
A joey just emerged to explore
A new world that vanished away
The boy Eli learning a language, the sore
Confusion tumbling Babel's tower down
A Zulu bride reciting her vows
Of eternal subservience, her life's crown
An old village doctor, the patients he saw
Slowly starve, the philosopher who unveiled
The reason for being, the wonder, the awe
A soldier's shadow cast on Archimedes

What will you cover black cloud?
Tall grass prairies buffaloes roamed
Horse-mounted ghosts, shedding blood
Tropical oceans shipwrecks fathomed
Failing to reach Eldorado's lure
Crystal white Alpine peaks Hannibal assailed
Challenging Rome's mighty endure
Vast highway networks which drained
The lifeblood of Earth, the tower
Of airfields and spaceports contraptions
Wombs of the doomsday machine's power

Who masterminded you bomb
Doctor Strangelove's pure pleasure
Faust's spectre that rose from the tomb
Brains amok digging Pandora's Box treasure
Or all of us, priests, servants and slaves
Metaphors' victims numbed by false flags
Aiming missiles at stars that can't save
From the lies, flying rags
Commit the godlike act, a clay nation
What right have you to shed a tear
For a burnt flower of creation

Abstract As these lines are written some 16,000 nuclear weapons are ready for launch worldwide. The details of a nuclear coup-de-grace may never be known. Once global communication channels break down, historians, should such survive, would not know exactly what took place—was it an accidental missile launch triggered by a computer glitch or by radar error, or a deliberate preventive strike following a use-them-or-lose-them strategy, where even a limited strike can trigger a near-total response. As cable and satellite communication cease and television, mobile phone and radio channels break down the hapless members of *H. sapiens,* having lived for more than 70 years under a nuclear Damocles Sword, would only be able to guess what has transpired. An attack on an urban population by 100 one-megaton weapons would kill up to 20% of the population through blast, heat, ground shock and neutrons and gamma rays, while an attack by 1000 one-megaton weapons would kill immediately almost half the population. Further deaths would occur from fires, lack of medical attention, starvation. Long-lived strontium-90 or cesium-137 would contaminate hot spots and enter the food chain causing internal damage. For survivors of a nuclear war, this lingering radiation hazard could represent a grave threat for as long as 1–5 years after the attack. Whereas models of nuclear war and nuclear winter, such as by Turco, Sagan and others have a large margin of error, the scale of a global or even large regional nuclear conflict can only lead to the largest calamity the human species has ever endured. Survivors of a nuclear event would emerge into an initially cold and subsequently further warming lands encroached by rising oceans.

3.1 Nuclear Winter

In 2016 the Bulletin of the Atomic Scientists stated: *Last year, the Science and Security Board moved the Doomsday Clock forward to three minutes to midnight, noting: 'The probability of global catastrophe is very high, and the actions needed to reduce the risks of disaster must be taken very soon.' That probability has not been reduced. The Clock ticks. Global danger looms. Wise leaders should act—immediately*[1] (Figs. 2.19 and 2.20).[2] Early in 2017 the arms of the clock were set to 2.5 min before midnight, the closest since 1953.

Nuclear war scenarios (Future of Life[3]) (Figs. 3.1 and 3.2) by accident or design (Starr 2015; Schlosser 2016) leave little doubt regarding the global effects of the release of hundreds or thousands of nuclear weapons. With the publication of *Nuclear Winter: Global Consequences of Multiple Nuclear Explosions* by Turco et al. (1983) the unthinkable horrors[4] have been exposed (NPR Books 2014), yet as these lines are written plans are announced for further development of nuclear weapons.[5]

Turco et al. (1983) use models previously developed to study the effects of volcanic eruptions in order to investigate attenuation of solar radiation by fine dust, smoke, soot and forest fires ignited by nuclear airbursts, leading to subfreezing land

[1] http://thebulletin.org/timeline

[2] http://thebulletin.org/timeline

[3] http://futureoflife.org/background/us-nuclear-targets/

[4] http://www.npr.org/2014/08/11/339131421/nuclear-command-and-control-a-history-of-false-alarms-and-near-catastrophes

[5] https://www.nytimes.com/interactive/2016/12/22/world/americas/trump-nuclear-tweet.html?_r=0

Fig. 3.1 Nuclear war systems: (1) The control room of the Savannah nuclear ship (https://commons.wikimedia.org/wiki/File:NS_Savannah_control_room_MD1.jpg#file); (2) A nuclear missile Silo of a Titan II ICBM in an underground complex. Steve Jurvetson. https://www.flickr.com/photos/jurvetson/7332367192. Creative Commons

Fig. 3.2 Stages in a nuclear exchange: from *left to right*: (1) nuclear explosions in the northern hemisphere; (2) the spread of soot and dust; (3) near total clouding by soot and dust (Paintings by Jon Lomberg, courtesy of Jon Lomberg. (Carl Sagan 1983). (4) snowy grave (https://pixabay.com/en/mountains-landscape-cross-1081799/. CCO Public Domain

temperatures (Turco et al. 1983; Ikle 2015). The explosion of nuclear warheads on the scale of thousands of megatons would create dust and smoke engulfing Earth within 1–2 weeks (Fig. 3.2), reducing sunlight to a few percent and land temperatures to −15° to −25°. Explosions on the scale of 100 megatons over major urban centers result in average hemispheric smoke of optical depths greater than 2 for weeks and, even in summer, subfreezing land temperatures for months. A 5000-megaton war at northern mid-latitude sites leads to radioactive fallout on time scales of days to weeks and to chronic mean doses of up to 500 mSv (50 rad) from gamma-ray exposure and to high internal radiation doses. Atmospheric disturbances caused by absorption of sunlight in smoke and dust clouds may greatly accelerate transport of particles and radioactivity from the Northern Hemisphere to the Southern Hemisphere. Subsequent enhancement of solar ultraviolet radiation due to ozone depletion, long-term exposure to cold, darkness and radioactivity could pose a serious threat to human survivors and other species (Turco et al. 1983).

The local and worldwide fallout effects of nuclear explosions depend on weapon design, explosive force, altitude and latitude of detonation, time of year, and local weather conditions.[6] For a 1 megaton bomb exploded at ground level in a 24 km/h wind the fallout extending hundreds of km downwind would induce lethal radiation of 6000 mSv at a distance of ~30–40 km. It is estimated an attack on U.S. population centers by 100 weapons of 1-megaton would kill up to 20% of the population immediately through blast, heat, ground shock and neutrons and gamma rays while an attack with 1000 1 Mt. weapons would kill immediately almost half the U.S. population with additional deaths from fires, lack of medical attention, starvation. Long-lived strontium-90 or cesium-137 would contaminate hot spots and enter the food chain causing internal damage. For survivors of a nuclear war, this lingering radiation hazard could represent a grave threat for as long as 1–5 years after the attack.[7]

A UN scientific committee estimated the cumulative per capita dose to the world's population as a result of atmospheric testing up to the year 2000 to be the equivalent of 2 years' exposure to natural background radiation, namely approximately 5 mSv. For most people the internal and external natural radiation amounts to less than 1 mSv annually. A nuclear war releasing 10 or 100 times the total yield of all previous weapons tests, namely ~50 to 500 mSv per-person, poses a global threat. Based on calculations by the US National Academy of Sciences the fallout from the 500-plus megatons of nuclear testing through 1970 will produce between 2 and 25 cases of genetic disease per million live births in the next generation. Thus between 3 and 50 persons per billion births will have genetic damage for each megaton of nuclear yield exploded and the induction of cancers range from 75 to 300 cases per megaton. Extrapolating these estimates to a nuclear war where 10,000 megatons are detonated, the effects on a world population of five billion over 30 years would include radiation-induced cancers and genetic damage to between 1.5 to 30 million people.[8]

According to an article titled *Nuclear War, Nuclear Winter, and Human Extinction* (Starr 2015).

1. A nuclear winter would cause most humans and large animals to die from nuclear famine in a mass extinction event similar to the one that wiped out the dinosaurs.
2. Nuclear firestorms would burn simultaneously over a total land surface area of many thousands or tens of thousands of square miles. These mass fires would release many tens of millions of tons of black carbon soot and smoke which would rise rapidly above cloud level and into the stratosphere.
3. Sunlight would heat the smoke, producing a self-lofting effect that would not only aid the rise of the smoke into the stratosphere but act to keep the smoke in the stratosphere for 10 years or more. Once in the stratosphere, the smoke would rapidly engulf the Earth and form a dense stratospheric smoke layer. The smoke from a war fought with strategic nuclear weapons would quickly prevent up to

[6] http://www.atomicarchive.com/Docs/Effects/wenw_chp2.shtml

[7] http://www.atomicarchive.com/Docs/Effects/wenw_chp2.shtml

[8] http://www.atomicarchive.com/Docs/Effects/wenw_chp2.shtml

70% of sunlight from reaching the surface of the Northern Hemisphere and 35% of sunlight from reaching the surface of the Southern Hemisphere.
4. Such an enormous loss of warming sunlight would produce Ice Age weather conditions on Earth in a matter of weeks. For a period of 1–3 years following the war, temperatures would fall below freezing every day in the central agricultural zones of North America and Eurasia.
5. Nuclear winter would cause average global surface temperatures to become colder than they were at the height of the last Ice Age. Such extreme cold would eliminate growing seasons for many years, probably for a decade or longer.
6. Temperatures would be much too cold to grow food, and they would remain this way long enough to cause most humans and animals to starve to death.
7. Global nuclear famine would result in a setting in which the infrastructure of the combatant nations has been totally destroyed, resulting in massive amounts of chemical and radioactive toxins being released into the biosphere.

Global leaders are presiding over an increasingly likely global suicide. The subjects of the catastrophic environmental consequences of nuclear war, and the increasingly likely activation of hair-trigger nuclear missile fleets by accident or design (Hallinan 2016), have been effectively omitted from major media outlets and from public debate.

The Executive Summary of a 1979 report on *The Effects of Nuclear War* submitted to the US Senate Committee on Foreign Affairs states[9]:

> *Nuclear war is not a comfortable subject. Throughout all the variations, possibilities, and uncertainties that this study describes, one theme is constant—a nuclear war would be a catastrophe. A militarily plausible nuclear attack, even "limited" could be expected to kill people and to inflict economic damage on a scale unprecedented in American experience; a large-scale nuclear exchange would be a calamity unprecedented in human history. The mind recoils from the effort to foresee the details of such a calamity, and from the careful explanation of the unavoidable uncertainties as to whether people would die from blast damage, from fallout radiation, or from starvation during the following winter. But the fact remains that nuclear war is possible, and the possibility of nuclear war has formed part of the foundation of international politics, and of U.S. policy, ever since nuclear weapons were used in 1945. The premise of this study is that those who deal with the large issues of world politics should understand what is known, and perhaps more importantly what is not known, about the likely consequences if efforts to deter and avoid nuclear war should fail. Those who deal with policy issues regarding nuclear weapons should know what such weapons can do, and the extent of the uncertainties about what such weapons might do.*

The major findings of this report[10] are summarized below:

The effects of a nuclear war that cannot be calculated are at least as important as those for which calculations are attempted. This is particularly true for indirect effects such as deaths resulting from injuries and the unavailability of medical care, or for economic damage resulting from disruption and disorganization rather than from direct destruction. In addition to the tens of millions of deaths during the days and weeks after the attack, there would probably be further millions (perhaps fur-

[9] http://atomicarchive.com/Docs/pdfs/7906.pdf

[10] http://atomicarchive.com/Docs/pdfs/7906.pdf

ther tens of millions) of deaths in the ensuing months or years. Nobody knows how to estimate the likelihood that industrial civilization might collapse in the areas attacked. The possibility of significant long-term ecological damage cannot be excluded. The impact of even a "small" or "limited" nuclear attack would be enormous. Predictions of the effects of such an attack are subject to the same uncertainties as predictions of the effects of an all-out attack. A review of calculations of the effects on civilian populations and economies of major counterforce attacks found that while the consequences might be endurable (since they would be on a scale with wars and epidemics that nations have endured in the past), the number of deaths might be as high as 20 million.

The desperate urgency of the launch-on-warning state of thousands of nuclear weapons has been grossly underestimated, receiving limited focus in the media. Sydney Drell, a physicist and nuclear weapons expert, commented on the proximity of a nuclear exchange in the following terms: Given all the close calls and mistakes in the 71 years since then (Hiroshima), he considered it a miracle that no other cities have been destroyed by a nuclear weapon—"*it is so far beyond my normal optimism*" (Schlosser 2016).

A corollary of the consequences of a regional to global nuclear exchange is furnished by the Marshall Islands tests, where a total of 6010 rad (60,100 mSv) were released over a period of 1 year[11] (Table 2.4).[12] Total surface radiation levels peaked in 1962 at 12,000 curies about 1965 and began to decline only in 1973 (Fig. 2.13). At this stage radiation levels reached 6.3×10^{9} Ci of ^{131}I.[13] Extrapolating from these tests to 500 Megaton, 1000 Megaton and 10,000 Megaton wars, ionizing radiation levels would reach 0.75×10^{22}Ci, 1.5×10^{22}Ci and 1.5×10^{23}Ci, respectively. As observed following the Marshall Islands nuclear tests, radiation would concentrate in mid-latitudes of the northern hemisphere (30–70°N) and will be lower in high northern latitudes (above 70°N). In the southern hemisphere radiation will be lower above 60°N.

The post-war damage could be as devastating as the damage from the actual nuclear explosion and the situation in which the survivors of a nuclear attack find themselves will be quite unprecedented. The collapse of agriculture in many parts of the world due to nuclear winter conditions over many months would result in widespread starvation. Major disruptions to food and fuel supplies, electricity, water and sewerage networks, medical services and pharmaceutical supplies could give rise to starvation and epidemics, killing more people than the nuclear blasts. Cities could disintegrate into survival units controlled by the military, police or armed gangs ravaging the countryside for food and fuel. The world would plunge into a new dark age (Fig. 3.3).

---

[11] http://www.deepspace.ucsb.edu/wp-content/uploads/2013/01/Effects-of-Nuclear-Weapons-1977-3rd-edition-complete.pdf

[12] https://en.wikipedia.org/wiki/Environmental_radioactivity#/media/File:US_fallout_exposure.png

[13] https://www.ctbto.org/nuclear-testing/the-effects-of-nuclear-testing/the-united-states-nuclear-testing-programme/

Fig. 3.3 The Scream, by Edvard Munch (https://pixabay.com/en/edvard-munch-scream-painting-terror-1332621/)

3.2 The Rising Oceans

Given CO_2 levels higher than 400 ppm and CO_{2e}(equivalent) levels ($CO_2 + CH_4 + N_xO$) higher than 450 ppm, sea level would continue to rise until they reach equilibrium with global temperatures, in turn a function of atmospheric greenhouse concentrations, including short or medium-term stadial cooling consequent on collapse of the Atlantic Meridional Ocean Circulation (AMOC) (Sect. 1.2.2).

The recent climate history of Earth allows projections of future trends based on atmospheric levels of greenhouse gases and aerosols and on the orbital relations between the Earth and the sun. Comparisons with intraglacial freezes (stadial) events; analogies with Pliocene to Miocene (2.56–23 Ma) climate conditions when mean global temperatures are estimated as ~2–4 °C above pre-industrial levels; Late Anthropocene rates of greenhouse gases and temperature rises; considerations regarding the Earth-sun Milankovic cycle; the habitat of plants; half-life spans of currently disseminated radioactive nuclides; the radiation-tolerance of organisms, and other factors constrain projections of future climates. Whereas the details and precise timing of future events can hardly be defined, overall trends consistent with these parameters may shed light on the nature of at least the next 20,000 years, the

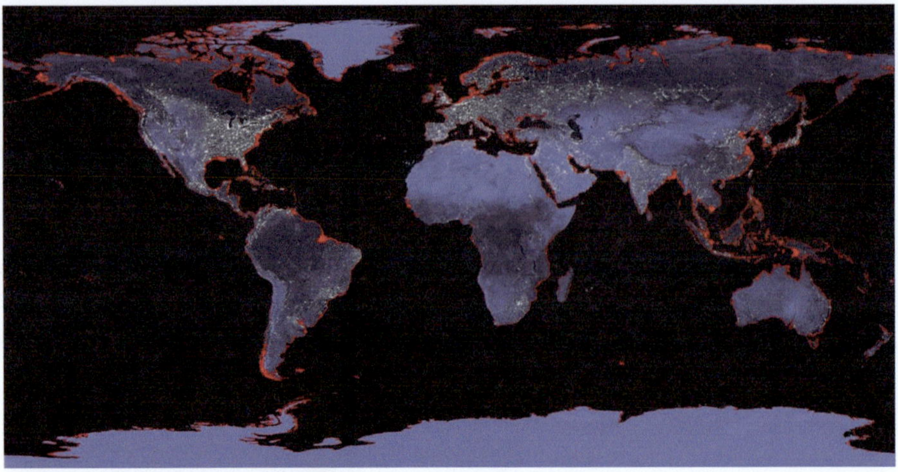

Fig. 3.4 Flood map for an assumed 6 m sea level rise (https://commons.wikimedia.org/wiki/File:6m_Sea_Level_Rise.jpg)

minimum age pending a descent to the next ice age (Berger and Loutre 2002) (Fig. 1.21) and the half-life span of plutonium[14] beyond the twenty-first century.

Early stages of the Plutocene would be accompanied by, or follow, a Late Anthropocene collapse of the Atlantic Meridional Overturning Circulation (AMOC), resulting in a stadial event,[15] indications of which are reported by Rahmstorf et al. (2015) in *Exceptional twentieth-century slowdown in Atlantic Ocean overturning circulation.* Such a stadial phase, possibly analogous to late Pleistocene *Younger dryas* (NOAA,[16] Steffensen et al. 2008) (12.900–11.700 BP), or to the early Holocene Laurentian glacial melt stadial event (Wagner et al. 2002) (8200 years BP), would result in extensive freeze conditions around the Atlantic Ocean.

The rise in sea level by near to 120 m (Lambeck et al. 2014) consequent on the melting of large parts of the Laurentian, Scandinavian, Greenland and Antarctic glaciers during the last glacial termination is resumed by the end of the Late Anthropocene, flooding lower river valleys—the cradles of civilization[17] and the hub of Neolithic and modern urban civilization. According to Hansen (2007) and Hansen et al. (2016) sea level rise on the scale of a few meters this Century is possible, stating "*We hypothesize that ice mass loss from the most vulnerable ice, sufficient to raise sea level several meters*" (Fig. 3.4). A sea level rise of 40 m above pre-industrial level, consistent with global temperatures about 4 °C above those the Late Anthropocene, would inundate long stretches of the Nile, Euphrates, Indus,

[14] https://en.wikipedia.org/wiki/Plutonium

[15] https://en.wikipedia.org/wiki/Stadial

[16] https://www.ncdc.noaa.gov/paleo/abrupt/data4.html

[17] https://www.khanacademy.org/humanities/ancient-art-civilizations/ancient-near-east1/the-ancient-near-east-an-introduction/a/the-cradle-of-civilization

Ganges, Brahmaputra, Mekong, Yellow River, Yangze, Mississippi, Amazon, Orinoco and other rivers, all shallow atoll islands in the Pacific and Indian oceans and numerous low-lying coastal zones around the world. Likewise some of the world's largest coastal and low river cities will be under salt water, including New York, Buenos Aires, London, Amsterdam, Hamburg, Copenhagen, Shanghai, Mumbai, Calcutta, Beijing, Jakarta, Ho Chi Minh City. Billions of people would have to escape to higher grounds, away from the fertile food-basket regions fertilized by freshwater rivers and lakes.

In the wake of the early Plutocene, in the aftermath of the abrupt runaway climatic effects and of radioactive contamination of soil and water on flora and fauna, environments may develop which bear significant analogies to those of the Pliocene and the Miocene when mean global temperatures ranged between 2 °C and 4 °C above pre-industrial stage of the Anthropocene, respectively. However, the extreme rise rates of greenhouse gases and of temperatures during the Late Anthropocene (Glikson 2016) and the associated mass extinction (Kolbert 2015) indicates the Plutocene would be highly depleted in higher order animals, opening habitats for proliferation of the Arthropods and of reptiles.

The Plutocene era emerges contemporaneously with the rise of global temperatures by more than one degree Celsius, leading to meters-scale sea level rise, interrupted by a collapse of the Atlantic Meridional Overturning Circulation (AMOC)[18] and a stadial freeze (Fig. 1.8). These climate developments may be accompanied by a rise in ionizing radiation, a destruction of the ozone layer and transient nuclear winter conditions induced by ash and soot. It would take more than 20,000 years for greenhouse gases to be sequestered into ocean and land, cooling the Earth to below 350 ppm CO_2.

James Hansen (2010) portrays a possible runaway greenhouse climate scenario referred to as the *Venus syndrome,* based on the theory that Venus (atmosphere—96.5% CO_2; 3.5% Nitrogen; surface temperature 462 °C) has undergone a greenhouse-driven heating in the past. An irreversible runaway greenhouse process on Earth raising the temperature of the atmosphere toward extreme levels is controversial.[19,20,21] Likely constraints are increased evaporation from the oceans under high temperatures and the effects of atmospheric water vapor, a greenhouse gas, on amplified heating of Earth. The balance between the cooling effects of evaporation, the warming effect of a water vapor-saturated atmosphere and the effects of precipitation under runaway greenhouse conditions are far from clear. There is no evidence from the paleo-climate record that the Earth has undergone a Venus syndrome, even when greenhouse gases and temperatures were abruptly elevated, for example following major volcanic events, asteroid impacts and methane release events (Glikson and Groves 2016). These events included (1) the extreme rises of

[18] https://usclivar.org/amoc

[19] https://arxiv.org/abs/1201.1593

[20] https://www.technologyreview.com/s/426608/how-likely-is-a-runaway-greenhouse-effect-on-earth/

[21] http://www.geoengineeringwatch.org/global-geoengineering-fueling-venus-syndrome/

CO_2 at the Permian-Triassic boundary event (<3400 ppm; Royer 2006); (2) end-Triassic (1300–2200 ppm; Beerling 2002, Beerling and Berner 2002); (3) K-T boundary impact: 350–500 ppm to at least 2300 ppm consistent with instantaneous transfer of ~4600 Gigaton Carbon (GtC) to the atmospheric reservoir sufficient to warm the Earth's surface by 7.5 °C (Beerling et al. 2002); (4) Paleocene-Eocene Thermal Maximum, elevating atmospheric CO_2 to near-1800 ppm at a rate of 0.18 ppm/year and mean temperatures rise of ~5 °C (Zachos et al. 2008). It is unlikely rises in temperature can reach such levels before human industry and transport systems can no longer emit CO_2 while global warming remains driven by amplifying feedbacks from land and oceans.

Projections of the early Plutocene depend on (1) the effects of a stadial freeze consequent on collapse of the AMOC, and (2) the timing of a nuclear exchange and nuclear winter. The ice core evidence indicates five stadial (freezing) events each following peak temperatures during interglacial phases of the late Pleistocene (Cortese et al. 2007). The time scale of the stadial may be estimated from comparisons with stadial phases such as the 8400–8100 years-old stadial triggered by Laurentian ice melt (Wagner et al. 2002; Barber et al. 1999) and the 12,900–11,700 years-old *Younger dryas* stadial.[22] However the time span of the imminent AMOC stadial is difficult to estimate in view of the extreme rise in CO_2 and temperature during the twenty to twenty-first centuries. The combination of a stadial freeze around the Atlantic Ocean and CO_2-driven temperature rises in other parts of Earth would result in unpredictable extreme pressure gradients[23] and abrupt pressure gradients between frozen regions and areas re-heated by high greenhouse gas concentrations, chaotic weather patterns, storms, collapse of agriculture in many parts of the world, widespread starvation and likely decline of the global populations to a fraction of the current 7.4 billion.[24]

The survival of humans during global warming, collapse of the AMOC (Tyrrell 2017), a nuclear war and its nuclear winter aftermath pose "*unknown unknowns*".[25] Predictions of the distribution and size of surviving groups are not possible. However, a number of factors pertain, including:

1. Collapse of agriculture in affected regions and partial to total breakdown of food supply, fuel transport, power lines, water supply and sewage systems, leading to partial to complete demise of urban centres dependent on these systems.
2. Flooding of coastal and low river agricultural regions and drying up of semi-arid food production regions would result in extensive famine.
3. A rise in infectious epidemics, such as the Spanish flu following WWI.
4. A rise in tropical and other diseases in the absence of antibiotics.

---

[22] https://www.ncbi.nlm.nih.gov/pubmed/18566247

[23] http://news.wisc.edu/abrupt-climate-change-could-follow-collapse-of-earths-oceanic-conveyor-belt/

[24] http://www.worldometers.info/world-population/

[25] https://www.brainyquote.com/quotes/quotes/d/donaldrums148142.html

5. Physical Wet-bulb warming effects of humid climates[26] during heat waves[27]
6. A breakdown of law and order leading to further disruption and cost lives

As these lines are written some 16,000 N-bombs are poised available for a suicidal "use them or lose them" strategy where perhaps some 50% may be dispatched, enough to kill billions. Much would depend on the sequence of events: following a short nuclear winter survivors would emerge into a heating world, or an AMOC collapse[28] taking place either before or after a nuclear event. The collapse as industry and transport systems will lead to a decline in CO_2 emissions. The increasing likelihood of global conflict and a consequent nuclear winter would constitute a coup-de-grace for industrial civilization (Carl Sagan[29]) though not necessarily for humans. Relatively low temperatures in the northern latitudes would allow humans to survive, in particular indigenous people genetically and culturally adapted to extreme conditions. Perhaps similar conditions could apply in high mountain valleys and elevated tropical islands. Vacated habitats would be occupied by burrowing animals and by radiation and temperature resistant Arthropods.

References

Barber DC et al (1999) Forcing of the cold event of 8,200 years ago by catastrophic drainage of Laurentide lakes. Nature 400:344–348
Beerling D (2002) CO2 and the end-Triassic mass extinction. Nature 415(6870):386–387
Beerling DJ, Berner RA (2002) Biogeochemical constraints on the Triassic-Jurassic boundary carbon cycle event. Glob Biogeochem Cycles 16(3):1036
Beerling DJ, Lomax BH, Royer DL, Upchurch GR, Kump LR (2002) An atmospheric pCO2 reconstruction across the cretaceous-tertiary boundary from leaf mega fossils. Proc Natl Acad Sci U S A 99(12):7836–7840
Berger A, Loutre MF (2002) An exceptionally long interglacial ahead? Science 297:1287–1288
Cortese G, Abelmann A, Gersonde A (2007) The last five glacial-interglacial transitions: a high resolution 450,000-year record from the subantarctic Atlantic. Paléo 22:PA4203
Glikson AY (2016) Cenozoic mean greenhouse gases and temperature changes with reference to the Anthropocene. Glob Chang Biol 22:3843–3858
Glikson AY, Groves C (2016) Climate, fire and human evolution: the deep time dimensions of the Anthropocene. Springer, Dordrecht, p 226
Hansen JE (2007) Scientific reticence and sea level rise. Environ Res Lett 2:024002
Hansen JE (2010) Storms of my grandchildren: the truth about the coming climate catastrophe and our last chance to save humanity. Bloomsbury, New York, p 303
Hansen JE et al (2016) Ice melt, sea level rise and superstorms: evidence from paleoclimate data, climate modeling, and modern observations that 2C global warming could be dangerous. Atmos Chem Phys 16:3761–3812

[26] https://en.wikipedia.org/wiki/Wet-bulb_temperature

[27] http://www.webmd.com/a-to-z-guides/heat-stroke-symptoms-and-treatment#1

[28] http://www.nature.com/nclimate/journal/v5/n5/full/nclimate2554.html

[29] http://www.e-reading.club/bookreader.php/148584/Sagan_-_The_Nuclear_Winter___The_World_After_Nuclear_War.pdf

Hallinan C (2016) We may be at a greater risk of nuclear catastrophe than during the cold war. Foreign Policy In Focus. http://fpif.org/may-greater-risk-nuclear-catastrophe-cold-war/

Ikle FC (2015) Worldwide Effects of Nuclear War. http://www.atomicarchive.com/Docs/Effects/wenw_foreword.shtml

Kolbert E (2015) The sixth extinction: an unnatural history. Henry Holt and Co, New York, p 307

Lambeck K et al (2014) Sea level and global ice volumes from the last glacial maximum to the Holocene. Proc Natl Acad Sci 111:15296–15303

NPR Books (2014) Nuclear 'Command And Control': a history of false alarms and near catastrophes. http://www.npr.org/2014/08/11/339131421/nuclear-command-and-control-a-history-offalse-alarms-and-near-catastrophes

Rahmstorf S et al (2015) Exceptional twentieth-century slowdown in Atlantic Ocean overturning circulation. Nat Clim Chang 5:475–480

Royer DL (2006) CO2-forced climate thresholds during the Phanerozoic. Geochim Cosmochim Acta 70:5665–5675

Sagan C (1983) The Nuclear winter: the World after Nuclear War. http://www.e-reading.club/bookreader.php/148584/Sagan_-_The_Nuclear_Winter___The_World_After_Nuclear_War.pdf

Schlosser E (2016) World war three by mistake: harsh political rhetoric, combined with the vulnerability of the nuclear command-and-control system, has made the risk of global catastrophe greater than ever. The New Yorker December 23, 2016. http://www.newyorker.com/news/news-desk/world-war-three-by-mistake

Starr S (2015) Nuclear War, Nuclear Winter, and Human Extinction. https://fas.org/pir-pubs/nuclear-war-nuclear-winter-and-human-extinction/

Steffensen JP et al (2008) High-resolution Greenland ice core data show abrupt climate change happens in few years. Science 321:680–684

Turco RP et al (1983) Nuclear winter: global consequences of multiple nuclear explosions. Science 222:1283–1292

Tyrrell KA (2017) Abrupt climate change could follow collapse of Earth's oceanic conveyor belt. University of Wisconsin–Madison News. http://news.wisc.edu/abrupt-climate-changecould-follow-collapse-of-earths-oceanic-conveyor-belt/

Wagner F, Aaby B, Visscher H (2002) Rapid atmospheric CO 2 changes associated with the 8,200-years-B.P. Cooling event. Proc Natl Acad Sci U S A 99:12011–12014

Zachos J, Dickens GR, Zeebe RE (2008) An early Cenozoic perspective on greenhouse warming and carbon-cycle dynamics. Nature 451:279–283

Chapter 4
A Republic of Insects and Grasses

Ozymandias
I met a traveler from an antique land
Who said: "Two vast and trunkless legs of stone
Stand in the desert. Near them on the sand,
Half sunk, a shattered visage lies, whose frown
And wrinkled lip and sneer of cold command
Tell that its sculptor well those passions read
Which yet survive, stamped on these lifeless things,
The hand that mocked them and the heart that fed.
And on the pedestal these words appear:
'My name is Ozymandias, King of Kings:
Look on my works, ye mighty, and despair!'
Nothing beside remains, round the decay
Of that colossal wreck, boundless and bare,
The lone and level sands stretch far away".

(Percy Bysshe Shelley)

Abstract While insights into the nature of the world in the wake of warming by several degrees Celsius and a nuclear war cannot be reached in detail, a number of projections arise from paleoclimate science and from current observations and trends. The increased concentrations of CO_2 above 405 ppm and CO_2e (equivalent CO_2 including methane) are tracking toward the stability threshold level of the Greenland and West Antarctic ice sheets and sea levels many meters higher than at present. Given sea levels in the range of 25 ± 12 m during the Pliocene pre-2.6 million years ago, the world's delta, low river valleys and coastal zones, the focus of much of human agriculture and civilization, will be flooded by sea water. As the Earth warms, the balance between increasing aridity in heated desert regions and enhanced hydrological cycle and precipitation in other regions would result in sharp climate gradients and intense storms. With a plutonium-239 half-life of approximately 20,000 years, the effects of radioactivity would decline. At this stage, large habitats vacated at the onset of the Plutocene due to climate tipping points and high

---

After Jonathan Schell's *Fate of the Earth*

A.Y. Glikson, *The Plutocene: Blueprints for a Post-Anthropocene Greenhouse Earth*, Modern Approaches in Solid Earth Sciences 13,
DOI 10.1007/978-3-319-57237-6_4

radioactivity would be re-occupied by tropical flora and fauna, notably the Arthropods. Accelerated speciation is observed during rebounds from mass extinctions and pulses of speciation appear sometimes to be associated with climate change.

4.1 Lessons from the Pliocene and Miocene

Comparisons of Plutocene temperatures with Cenozoic environments under global temperatures estimated at approximately 2– 4 °C higher than Late Holocene pre-industrial temperatures are potentially instructive, with an essential proviso. As distinct from the gradual evolution of Miocene and Pliocene climates and habitats, post-Anthropocene climates follow an abrupt shift in state and tipping points reached within less than a century, involving a mass extinction of terrestrial and marine fauna.

The current CO_2 level of 405 ppm and higher than 450 CO_2e (equivalent of CO_2 + methane) are already within the estimated range of the Pliocene (5.333 to 2.58 Ma[1]) as based on proxy data as 360 to 400 ppm,[2] with mean global temperatures of 2–3 °C above pre-industrial temperatures. Mid-Pliocene sea levels range from 10 to 40 m above present, with a mean value of 25 ± 12 m, indicating complete deglaciation of the West Antarctic and Greenland ice sheets, and significant loss of mass in the East Antarctic ice sheet.[3,4,5] Palaeoclimate proxies indicate overall higher precipitation and lower aridity. The observation emerges of a tropical to semi-tropical Earth with smaller ice caps and many of the present delta and lower river zones covered by sea water.

The Miocene (23.03 to 5.333 million years ago) is characterized by a series of key climatic events that led to the founding of the late Cenozoic glacial climates and the dawn of modern biota. Kurschner et al. (2007) present a CO_2 record derived from stomatal frequency data of tree species, displaying marked CO_2 fluctuations in the range of 300–600 ppm where periods of low CO_2 coincide with glaciations while periods of elevated CO_2 of 500 ppm coincide with warm climatic optimum (Fig. 4.1). Major changes in Miocene terrestrial ecosystems including expansion of grasslands and radiations among terrestrial herbivores are linked to marked fluctuations in CO_2. The Miocene involved an initial 21–16 Ma warm to moderately warm period followed by slow cooling with fire-prone forests dwindling in extent relatively

[1] https://en.wikipedia.org/wiki/Pliocene

[2] http://www.nature.com/ngeo/journal/v4/n5/full/ngeo1118.html

[3] http://www.nature.com/ngeo/journal/v4/n5/full/ngeo1118.html

[4] http://www.sciencedirect.com/science/article/pii/0921818194900167

[5] https://www.ipcc.ch/publications_and_data/ar4/wg1/en/ch6s6-3-2.html

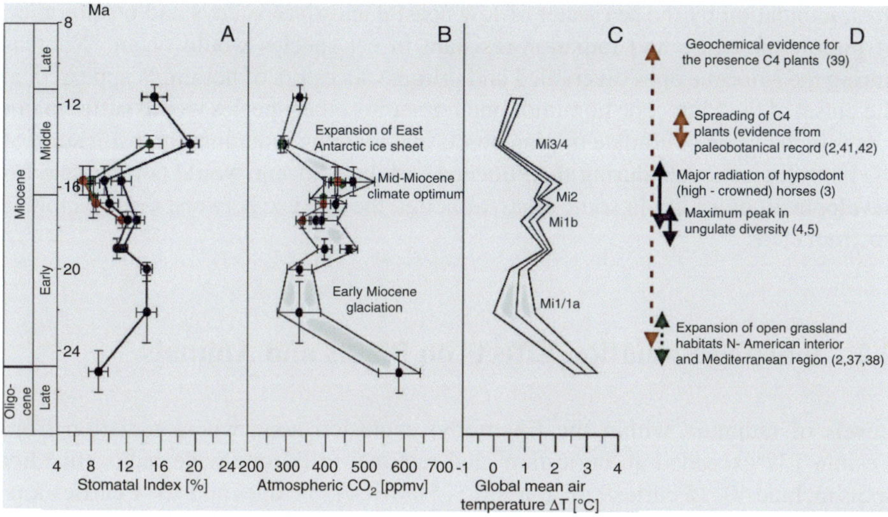

Fig. 4.1 Late Oligocene–Miocene stomatal index records, inferred atmospheric CO_2 fluctuations, and effects on global temperature compared with developments in terrestrial ecosystems (Kurschner et al. 2007) (By Permission by Proc. Nat. Acad. Sci.)

to expanding fire-resistant grassland[6] allowing herbivore animals such as horses, rhinoceroses, and hippos to proliferate along with their carnivorous pursuers. About 95% of modern plants already existed by the end of the Miocene epoch. During warm periods high temperatures and atmospheric energy were bound to trigger cyclones, heat waves and fires. In the latter respect Miocene climate and Plutocene climate would share important analogies. By the middle Miocene at 14.8–14.5 Ma ago (Langhian stage) extinctions of terrestrial and aquatic life forms were associated with an abrupt cooling event. By 8 Ma temperatures dropped sharply, the Antarctic ice sheet extended to near present-day size, while the climate remained warm supporting forests well into the Pliocene.[7] Miocene plants, mammals and birds were well-established, including whales, seals, and kelp spread.[8]

In so far as mean global temperatures are projected to rise by about 4 °C late in the twenty-first century[9,10,11] extensive loss of forests due to fast rise of temperatures,

[6]https://en.wikipedia.org/wiki/Miocene

[7]https://en.wikipedia.org/wiki/Miocene

[8]https://en.wikipedia.org/wiki/Miocene

[9]https://www.ipcc.ch/publications_and_data/ar4/wg1/en/spmsspm-projections-of.html

[10]http://www.climatechangenews.com/2014/04/01/how-much-worse-is-a-4-degrees-world/

[11]Global Warming, our future. http://globalwarming.berrens.nl/globalwarming.htm

fires, inundation by the sea water of low coastal and river valleys and proliferation of high-temperature and radiation-resistant insect species would occur. Whereas during the Miocene apes diversified and primate ancestors of hominids appeared, at the outset of the Plutocene hominids and numerous other species would suffer major losses in the wake of climate tipping points and ionizing radiation. Intensification of the hydrological cycle during the Pliocene and the Miocene would compensate for development of aridity in some areas, although the balance between these factors is far from clear.

4.2 Ionizing Radiation Effects on Plants and Animals

Levels of radiation within the Chernobyl exclusion zone where radiation from Cesium-137 exceeded 40 curies/km^2 and regional to Europe-wide radioactive hot spots include 40–15 curies/km^2 zones, 15–5 curries/km^2 zones and 15–1 curries/km^2 zones (Figs. 2.21 and 2.22). In terms of cumulative radiation levels, areas around Chernobyl are classified as 3700–1480 × 10^3 Bq, 1480–555 × 10^3 Bq, 555–185 × 10^3 Bq and 185–37 × 10^3 Bq (Fig. 2.21). The extreme radiation levels of the Chernobyl exclusion zones may be compared with the cumulative radiation scales reached in some of the areas affected by the nuclear weapon tests including the Castle Bravo hydrogen bomb test (Fig. 2.8)[12] with a downwind radiation gradient from 7500 curies to few hundred curies levels. Whereas effects on forests around Chernobyl are evident (Fig. 4.2[13]), given fatal radiation doses in the range of 4–12 Sievert for mammals (Table 4.2) a high level of survival of flora and fauna is observed within the Chernobyl exclusion zone (Deryabina et al. 2015).

According to Deryabina et al. (2015) "*Several previous studies of the Chernobyl exclusion zone indicated major radiation effects and pronounced reductions in wildlife populations at dose rates well below those thought to cause significant impacts. In contrast, our long-term empirical data showed no evidence of a negative influence of radiation on mammal abundance. Relative abundances of elk, roe deer, red deer and wild boar within the Chernobyl exclusion zone are similar to those in four (uncontaminated) nature reserves in the region and wolf abundance is more than 7 times higher. Additionally, our earlier helicopter survey data show rising trends in elk, roe deer and wild boar abundances from one to ten years post-accident. These results demonstrate for the first time that, regardless of potential radiation effects on individual animals, the Chernobyl exclusion zone supports an abundant mammal community after nearly three decades of chronic radiation exposures.*"

---

[12] https://en.wikipedia.org/wiki/Castle_Bravo

[13] https://en.wikipedia.org/wiki/Chernobyl_disaster#Flora_and_fauna

Fig. 4.2 Following the Chernobyl disaster four square kilometers of pine forest directly downwind of the reactor turned *reddish-brown* and died, earning the name of the *Red Forest* (https://en.wikipedia.org/wiki/Chernobyl_disaster#Flora_and_fauna)

However, significant genetic changes have been identified within the exclusion zone around Chernobyl where experimental bean plants display reduction in weight and water uptake and contain more heavy metals-binding proteins and compounds found to reduce chromosomal abnormalities in human blood exposed to radiation.[14]

[14] http://www.sciencemag.org/news/2009/05/how-plants-survived-chernobyl

4.3 Implications of Nuclear Tests for the Plutocene

The highly uneven distribution of radioactive fallout world-wide, controlled by winds and air-turbulence cells results in radiation hot spots, as in the Nevada tests (Fig. 2.4) and Chernobyl accident (Figs. 2.21 and 2.22). Large parts of the radioactive fallout are washed to rivers and the sea, where long lived plutonium accumulates in abyssal clay layers, constituting records of the Late Anthropocene and the Plutocene. The heterogeneous distribution of radioactive fallout results in extreme radiation regime in many areas, oxidation of atmospheric nitrogen, destruction of the ozone layer and associated clouding by aerosols and carbon soot, yielding a clue to the unthinkable consequences of a multi-megaton nuclear war.

Estimates of radioactive fallout triggered by a nuclear exchange are complicated by numerous factors, including the yield of nuclear weapons, the nature of the targets, the geographic areas and levels below ground, above ground or undersea at which they are detonated, local weather including winds and rain, ocean currents and other factors (Glasstone and Dolan (1950/1977).[15,16,17] Radioactivity would be greatly enhanced should nuclear power plants be targeted, with consequent dissemination of ionizing radiation regionally to world-wide (Martin 1988).[18] Table 4.1 lists estimates of the order of magnitude of released radioactivity during a nuclear war corresponding to explosive power of 500, 1000 and 10,000 megaton TNT, based on estimates of the ionizing radiation of exploded nuclear weapons from earlier nuclear tests on scales ranging from 1 kiloton[19] to 15 megatons.[20,21] Table 4.2 lists the lethal doses of radiation on animals. It is clear that, of all living beings, it is the arthropods that are best suited to withstand the effects of ionizing radiation.

[15] https://nnsa.energy.gov/sites/default/files/nnsa/inlinefiles/glasstone%20and%20dolan%201977.pdf

[16] https://www3.nd.edu/~nsl/Lectures/phys205/pdf/Nuclear_Warfare_9.pdf

[17] https://www.un.org/disarmament/wmd/nuclear/

[18] Martin (1988) Nuclear winter: science and politics and Martin (2016)

[19] http://www.deepspace.ucsb.edu/wp-content/uploads/2013/01/Effects-of-Nuclear-Weapons-1977-3rd-edition-complete.pdf

[20] https://en.wikipedia.org/wiki/Castle_Bravo#Fallout

[21] https://www3.nd.edu/~nsl/Lectures/phys205/pdf/Nuclear_Warfare_9.pdf

Table 4.1 Explosive intensity (in kilotons and megatons) and radiation values (in curies and Becquerel) of nuclear tests and global fallout values, with extrapolations to possible nuclear war levels of fallout levels per square km and per capita for a world population of seven billion people, compared to natural radiation level

	Explosion/fallout	Explosive power	N-Tests Radiation (1 Curie (Ci) = 3.7×10^{10} Becquerel (Bq)	Extrapolated Radiation per 500 Mt	Radiation per 1000 Mt	Radiation per 10,000 Mt
A	Natural radiation per capita[a,b]		0.2 μCi (^{40}K, ^{14}C) = 0.74×10^4Bq (~7400 Bq)			
B	One Kiloton Atomic Test[c]	1 Kt	3.10^{10} Ci = 1.1×10^{21}Bq	1.5×10^{16} Ci	3×10^{16} Ci	3×10^{17} Ci
C	Bravo H-bomb test[d]	15 Mt	6.3×10^9 Ci^{131}I = 2.3*10^{20}Bq	2.1×10^{11}Ci ^{131}I = 7.8×10^{21}Bq	4.2×10^{11} Ci = 1.56×10^{22}Bq	4.2×10^{12} Ci = 1.56×10^{23}Bq
D	Global radiation from N-weapon tests and SNAP-9A[e]	~500 Mt	3.46×10^5Ci	3.46×10^5Ci	6.92×10^5Ci	6.92×10^6Ci
			1.28×10^{16}Bq	1.28×10^{16}Bq	2.56×10^{16}Bq	2.56×10^{17}Bq
E	fallout Northern Hemisphere (N-test and SNAP-9A)		2.56×10^5Ci			
			9.4×10^{15}Bq			
F	Pu 239, 240 fallout Southern Hemisphere (N-test and SNAP-9A)		6.9×10^4Ci			
			2.5×10^{15}Bq			
G	Potential radiation of the Modern arsenal (Extrapolated from B)			0.75×10^{22}Ci	1.5×10^{22}Ci	1.5×10^{23}Ci

(continued)

Table 4.1 (continued)

	Explosion/ fallout	Explosive power	N-Tests Radiation (1 Curie (Ci) = 3.7×10^{10} Becquerel (Bq)	Extrapolated Radiation per 500 Mt	Radiation per 1000 Mt	Radiation per 10,000 Mt
H	Radiation per km^2 ($5.1 \times 10^8 km^2$) (from D)		$3.46 \times 10^5 Ci/5.1 \times 10^8 km^2 = 0.68 \times 10^{-3} Ci/km^2 = 2.5 \times 10^7 Bq/km^2$			
	Fallout on continents 30% global)			$0.8 \times 10^7 Bq/km^2$	$1.6 \times 10^7 Bq/km^2$	$16 \times 10^7 Bq/km^2$
I	Radiation per capita (for world population of seven billion) (from D)		$3.46 \times 10^4 Ci/7 \times 10^9 = 0.49 \times 10^{-5}$ Ci/per capita = 1.8×10^5 Bq/per capita			

Radiation level for the Castle Bravo test is the maximum indicated in Fig. 2.13. Note: the extreme fallout values constitute averages—the geographic distribution of explosions and the weather would result in highly differential radiation regimes

[a]https://en.wikipedia.org/wiki/Curie The typical human body contains roughly 0.1 μCi (14 mg) of naturally occurring Potassium-40. A human body containing 16 kg of Carbon (see composition of the human body) would also have about 24 nanograms or 0.1 μCi of Carbon-14. Together, these would result in a total of approximately 0.2 μCi or 7400 Bq inside the person's body. (μCi) is 0.000001 curie

[b]http://www.world-nuclear.org/information-library/safety-and-security/radiation-and-health/radiation-and-life.aspx

[c]http://www.deepspace.ucsb.edu/wp-content/uploads/2013/01/Effects-of-Nuclear-Weapons-1977-3rd-edition-complete.pdf

[d]https://www.ctbto.org/nuclear-testing/the-effects-of-nuclear-testing/the-united-states-nuclear-testing-programme/. According to the U.S. Centers for Disease Control and Prevention, the tests in the Marshall Islands released 6.3 billion curies (See Chart 2) of I-131 alone, which is 42 times the amount released by atmospheric nuclear testing in Nevada

[e]http://www.nature.com/nature/journal/v241/n5390/abs/241444a0.html

Table 4.2 Lethal radiation doses on animals (in Sieverts)[a]

Organism	Class/Kingdom	Lethal dose (Sv)	LD_{50} (Sv)	LD_{100} (Sv)
Dog	Mammals		3.5 ($LD_{50/30\ days}$)[b]	
Human	Mammals	4–10	4.5	10
Rat	Mammals		7.5	
Mouse	Mammals	4.5–12	8.6–9	
Rabbit	Mammals		8 ($LD_{50/30\ days}$)	
Tortoise	Reptile		15 ($LD_{50/30\ days}$)	
Goldfish	Fish		20 ($LD_{50/30\ days}$)	
Escherichia coli	Bacteria	60		60
German cockroach	Insects		64	
Shellfish	–		200 ($LD_{50/30\ days}$)	
Fruit fly	Insects	640		
C. elegans	Nematode		160–200	$\gg$ 500–800
Amoeba	–		1000 ($LD_{50/30\ days}$)	
Braconidae	Insects	1800		
Milnesium tardigradum	Eutardigrade	5000		
Deinococcus radiodurans	Bacteria	15,000		
Thermococcus gammatolerans	Archaea	30,000		

[a]https://en.wikipedia.org/wiki/Radioresistance
[b]A dose killing 50% of people within 30 days

4.4 Arthropod Civilizations

According to Sepkoski (1998) an acceleration of speciation during rebounds from mass extinctions is observed in the paleontological record, pulses of speciation appear sometimes to be associated with climate change.[22] The colonial insects—bees, wasps, ants and termites, construct sophisticated civilizations that withstand close analogies with and in some respects surpass human civilizations. An example are the termites, the most successful insects on Earth, colonizing most landmasses except for Antarctica,[23] building immense structures as compared with the size of individual termites, with remarkable complexity and coherence, which serve as wind-driven lungs for the colony (Fig. 4.3a–d).[24] Termites, whose ancestors may have appeared in the Permian more than 250 million years ago (Tilyard 1937) have survived extreme climates and several mass extinction events and have similar radiation resistance as cockroaches to which they are related (Turner 2015). Originally living in the rainforest, fungus farming allowed termites to colonize the African

[22] https://www.ncbi.nlm.nih.gov/pmc/articles/PMC1692211/pdf/11541734.pdf
[23] https://en.wikipedia.org/wiki/Termite
[24] http://www.esf.edu/efb/turner/turner%202011/publication%20pdfs/Termites%20as%20models%20of%20swarm%20cognition%20personal%20copy.pdf

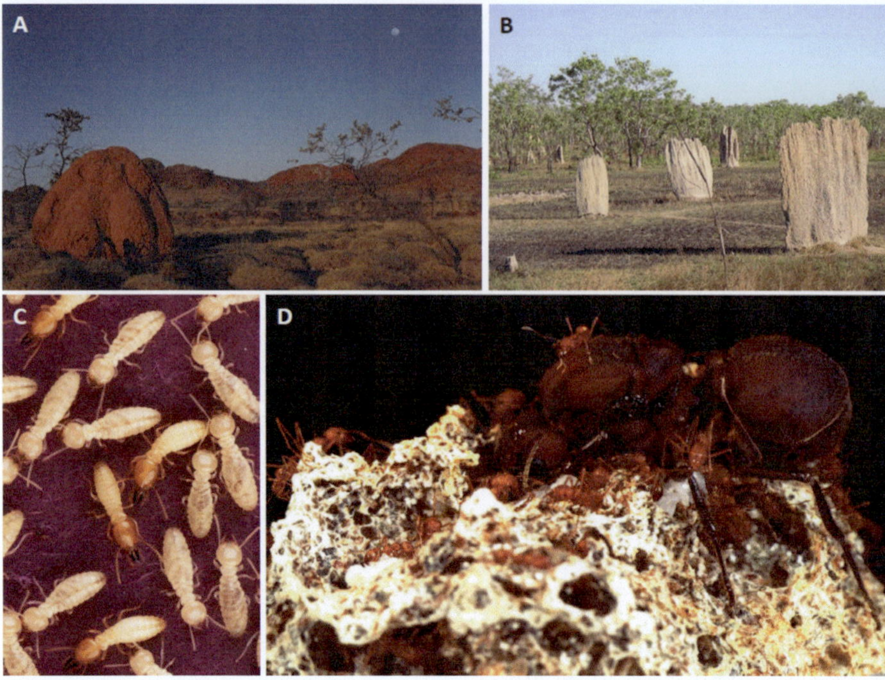

Fig. 4.3 (**a**) Termite nests, north-central Pilbara, northern Western Australia; (**b**) Termite nests, Litchfield National Park, Northern Territory, Australia (https://en.wikipedia.org/wiki/Litchfield_National_Park#/media/File:Litchfield_National_Park-Termite_mounds.jpg); (**c**) Formosan subterranean termite (Coptotermes formosanus). Soldiers (*red-colored heads*), Workers (*pale-coloured heads*) (https://en.wikipedia.org/wiki/Termite); (**d**) Atta columbica queen with workers on fungus substrate (https://commons.wikimedia.org/wiki/File:Atta_colombica_queen.jpg)

savannah and other new environments, eventually expanding into Asia (Roberts et al. 2016). The thermo-tolerant nature of ant species such as *Cataglyphis, Ocymyrmex, and Melophorus*, from the Sahara, Namib and Australian Deserts, respectively, are known to forage in surface temperatures above 60 °C (Sherwood 1996), which would allow them to survive under global warming conditions of 4 °C and higher above pre-industrial temperatures.

Subterranean termites protect themselves from temperature extremes through the construction of tubular air conditioning systems, capturing outside wind, mixing it with stale inner air and venting out excess heat, thus keeping the nest cool and dry. Termites foraging for food above ground protect themselves with shelter tubes built from particles of soil or wood and bits of debris held together by salivary secretions. The tubes may be thinly constructed or large and thick-walled to accommodate many termites moving vertically between the soil and the food source.[25] Moisture is important to subterranean termites, which have very little resistance to dehydration.

[25] https://www.termatrac.com/home-owners/learn-about-termites/termite-biology/

To survive, they must maintain contact with the soil and above-ground moisture sources. The extent to which termites can withstand high temperatures is demonstrated by their survival in microwave ovens.[26] The radiation resistance of insects, such as fruit flies (<640 Sievert) and wasps (Braconidae <1800 Sievert) would allow them to survive and thrive in environments vacated by, mammals and reptiles (Table 4.2). Through the spread of termite nests and the underground cultivation of fungus from foraged vegetable matter arid soils become re-fertilized allowing the bush to thrive.

The collective intelligence and swarm paradigm of insect colonies constitute frontier fields of research in entomology (David Gordon[27]). Based on detailed observations of communications and social and work relations among termites as well as experiments on termite nests in Namibia Scott Turner (2011, 2015) states: "*These are challenging tasks for tiny insects individually, but collectively they seem to solve them just fine, even when the repair extends far beyond the six-week lifespan of a typical termite worker*" and "*The collective intelligence of the colony is quite real, as real as our own intelligence, and we are far from comprehending either.*" Further observations may suggest animal-like and to a significant extent human-like, rather than robotic behaviour: "*Eventually, it hit me: these are not robots; they are living things with individuality, wants and desires. A robot cannot ever "want" to be groomed or "want" to give water to another or "want" a drink. But termites seemingly do. And this gives termites, both individually and collectively, something like a soul – an animating principle that one does not find in mere machines. It needn't be some vital "stuff" as the ancients once thought, but still something ineffable that makes life distinct from nonlife.*" The swarm's cognitive abilities arise from both interaction amongst the individual agents within a swarm, and from the interaction of the swarm with the environment, mediated by the mound's dynamic architecture.

The social structure of termite, ant and wasp colonies has been likened to a *super organism*, in which the colony has many of the attributes of a more conventionally defined organism, including physiological and structural differentiation and coordinated and goal-directed action. In so far as cognition constitutes a social phenomenon insect colonies should be imbued with cognition, whether units of the social system are cells or neurons in brains, as among mammals, or organisms in societies, as are colonial insects (Turner 2011). Eugene Marais (1872–1936) made groundbreaking studies into termites, describing the analogies between work processes inside termite society and the workings of the human body (Marais 1937/2009). Marais regarded red and white soldiers as analogous to blood cells, the queen as the brain and the termites' mating flight in which individuals from separate termitaries leave to produce new colonies as exactly equivalent to the movement of sperm and ova. The colony super-organism possesses cognitive powers allowing it to respond adaptively to environmental contexts and problems beyond the ability of individuals within the swarm (Bonabeau et al. 1999).

---

[26] https://www.quora.com/How-do-ants-incredibly-survive-even-after-cooking-in-microwave-oven

[27] http://ai-depot.com/Essay/SocialInsects.html

References

Bonabeau E et al (1999) Swarm intelligence: from natural to artificial systems. Oxford University Press, Santa Fe Institute Studies in the Sciences of Complexity, JASSS. http://jasss.soc.surrey.ac.uk/4/1/reviews/kluegl.html

Deryabina TG et al (2015) Long-term census data reveal abundant wildlife populations at Chernobyl. Curr Biol 25(19):R824–R826

Glasstone S, Dolan PJ (1950/1977) The effects of nuclear weapons. US Dept Defense and Energy Research and Development Administration. http://www.deepspace.ucsb.edu/wp-content/uploads/2013/01/Effects-of-Nuclear-Weapons-1977-3rd-edition-complete.pdf

Kurschner WM et al (2007) The impact of Miocene atmospheric carbon dioxide fluctuations on climate and the evolution of terrestrial ecosystems. Proc Natl Acad Sci 105(2):449–453

Marais EN (1937/2009) The soul of the white ant. Osiran Books, Pretortia, p 168

Martin B (1988) Nuclear winter: science and politics. Science and Public Policy, Vol. 15, 321–334. http://www.bmartin.cc/pubs/88spp.html

Martin D (2016) Risk of nuclear attack rises. http://www.cbsnews.com/news/60-minutes-risk-of-nuclear-attack-rises/

Roberts EM et al (2016) Oligocene termite nests with in situ fungus gardens from the Rukwa Rift Basin, Tanzania, support a Paleogene African origin for insect agriculture. Plos One. http://journals.plos.org/plosone/article?id=10.1371/journal.pone.0156847

Sepkoski JJ (1998) Rates of speciation in the fossil record. Phil Trans R Soc Lond B353:315–326

Sherwood V (1996) Book of insects records. University of Florida, Gainesville. Chapter 1: Most heat tolerant

Tilyard RJ (1937) Kansas Permian insects. Part XX the cockroaches, or order Blattaria. Am J Sci 34:169–202. (249–276)

Turner JS (2011) Termites as models of swarm cognition. Swarm Intell 5:19–43

Turner JS (2015) The 'Collective Mind' of the Termite. Live Science October 30, 2015. http://www.livescience.com/52644-the-collective-mind-of-the-termite.html

Chapter 5
Homo Prometheus

Stars fell to the Earth in a blaze of light
And where they fell monsters were born,
Hideous and blind.

(Gore Vidal, in Messiah)

Falling Star
For an infinitely long second nascent
Your tail plotted a crescent, a fiery arc
From a star-spangled sky incandescent
To the fast sleeping Earth in the dark
Your grave for all time

Who are you, friend or foe, stranger
A fragment fallen by a space highway
Of the asteroid belt, posing danger
Or some orbit-decayed fancy hardware
Of fatal star wars fleets

Are you a harbinger of good news
Or signal dire distress
For this embattled Earth in blues
Will your fleeting torch impress
On my race—a new truth

Will you plunge way beyond yonder
Or fall here, close by my side
On this red desert dune, I wonder
Why I am stranded tonight, wide eyed
In awe, with no faith

A.Y. Glikson, *The Plutocene: Blueprints for a Post-Anthropocene Greenhouse Earth*, Modern Approaches in Solid Earth Sciences 13,
DOI 10.1007/978-3-319-57237-6_5

Abstract The human journey, emerging at least 2 million years ago with the appearance of *Homo erectus* and the mastery of fire, has become an inexplicable phenomenon in natural history, not least because it is the distant descendants of this species who are confronted with the ultimate consequences of its evolution. It took *H. erectus,* physically endowed hunters, at least another million years before signs of ritual burial appeared with *Homo Heidelbergensis*. Not until about 160,000 years-ago does evidence for incipient art emerge. While records of farming emerge 23,000 years ago, it is only 12,000 years ago that more extensive agriculture is recorded in Middle East river valleys. With new food resources and wealth the human mind turned from camp-fire rituals to the construction of monuments for the dead and to mass killings named "war", which dominate the history of civilization. From wars motivated by a need for resources deadly conflicts under a myriad of false flags have become an end for itself, including the most bizarre forms of blood rituals and torture. Since the industrial age with the development of new weapons for industrial-scale killing, and since the splitting of the atom, "war" has become a form of mass suicide threatening to destroy civilization and much of nature, leading to the 7th mass extinction of species. It would appear, in the sense of the Yin and Yen, the species highest forms of art and science and horrendous atrocities are in some mysterious way intertwined.

5.1 Homo Prometheus

An understanding of the origins of the Seventh mass extinction of species, if ever to be achieved, requires possible corollaries related to Darwinian evolution theory or, by contrast, Cuvier catastrophism. The polarity between gradualist uniformitarian theories of evolution (James Hutton 1726–1797[1]; Charles Lyell 1797–1875[2]) and catastrophism (Georges Cuvier 1769–1832[3]) has been progressively defined through advances in the study of paleontology, sedimentation, paleo-climate research, and volcanism and asteroid impacts, accompanied by precise isotopic age determinations. These studies identify periods of gradual evolution interrupted by abrupt events at rates to which many species, and in some instances much of the biosphere, could not adapt, with consequent large scale extinctions. Detailed investigations of the carbon, oxygen and sulphur cycles using a range of proxies, including leaf pore stomata, $\delta^{13}C$, $\delta^{34}S$ and $^{87/86}Sr$ isotopes, as well as geochemical mass balance modeling, provide evidence of major evolutionary trends as well as distinct events during the Phanerozoic (542 Ma to the present). These studies identify alternating atmospheric greenhouse periods with CO_2 levels of ~2000–5000 ppm and glacial phases

[1] http://www.amnh.org/explore/resource-collections/earth-inside-and-out/james-hutton-the-founder-of-modern-geology/

[2] http://www.famousscientists.org/charles-lyell/

[3] http://www.ucmp.berkeley.edu/history/cuvier.html

with CO_2 levels of <500 ppm, exerting major controls over biological evolution (Glikson and Groves 2016).

The geological record betrays a close correspondence between paleontological, sedimentary, volcanic, asteroid impact, paleo-CO_2 and paleo-temperature trends, identifying environmental factors underlying the evolution and extinction of species (McElwain and Punyasena 2007[4]; Beerling 2002; Keller 2005; Glikson 2005; Kolbert 2015; Stanley 2016 (Fig. 1.2). High-resolution regional palaeoecological studies indicate extensive ecological upheavals, high species-level turnover and recovery intervals lasting up to millions of years, with close correlations with transformation of terrestrial vegetation. In the Late Anthropocene, based on comprehensive geological studies of deep ocean sediment cores, coral reefs, distribution patterns of forests and variations in the distribution of communities, numerous species are facing extinction at a rate reaching between 1000 to 10,000 times the previous natural rates[5,6,7] Given the longevity of greenhouse gases, the current mass extinction is tracking toward devastation of the biosphere on a scale commensurate with mass extinction events triggered by volcanic and asteroid impact events (Fig. 1.2).

The mastery of fire in the mid-Pleistocene about ~2.0–1.0 Ma coincides with accentuation of intermittent glacial conditions associated with the amplification of the 41 kyr-long Milankovic cycles about 1.8–1.5 Ma-ago, a period of increased climate variability (Fig. 5.1). The appearance of a species which has learnt to kindle fire (Wrangham 2009) (Fig. 5.2) meant that, for the first time, the flammable carbon-rich biosphere could be ignited by a living organism. It is likely that, as is the case with major inventions, the mastery of fire was driven by necessity under the acute environmental pressures associated with the descent from the warm Pliocene (5.2–2.6 Ma) climate into the abrupt glacial-interglacial cycles of the Pleistocene (2.6–0.01 Ma), with global mean temperature oscillations of up to ±5 °C (Fig. 5.1) (deMenocal 2004; Glikson and Groves 2016).

The effects of fire on the human mind are known to those who have spent long periods around campfires, as the author has on-and-off for 40 years. For hundreds of thousands of years, gathered during the long nights around camp fires, captivated by the flickering life-like dance of the flames, humans developed imagination, insights, cravings, fears, premonitions of death and thereby aspirations for immortality, omniscience, omnipotence and concepts of supernatural gods. Pantheistic beliefs revered the Earth, its rocks and living creatures. Fear, an instinctive sense arising in animals when endangered, is created in the human mind in advance, allowing it to foresee future threats.[8] Since at least 160,000 years-ago although possibly some 500,000 years ago that a perception of death and thereby a yearning for immortality

---

[4] https://www.ncbi.nlm.nih.gov/pubmed/17919771

[5] http://cmsdata.iucn.org/downloads/species_extinction_05_2007.pdf

[6] http://www.biologicaldiversity.org/programs/biodiversity/elements_of_biodiversity/extinction_crisis/

[7] http://wwf.panda.org/about_our_earth/biodiversity/biodiversity/

[8] http://www.cerebromente.org.br/n05/doencas/fobias_i.htm

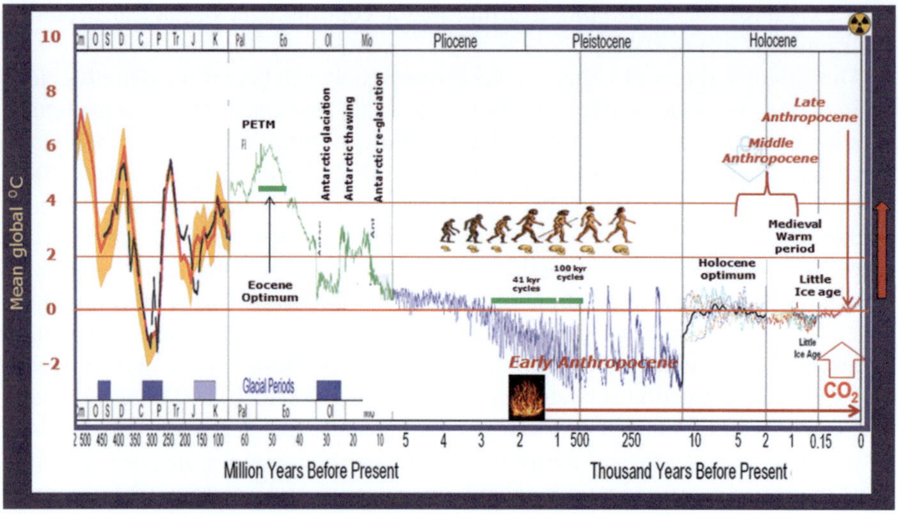

Fig. 5.1 Geological and anthropological evolution through the Phanerozoic

Fig. 5.2 A Diorama showing ancient cavemen lighting fire, National Museum of Mongolian History in Ulaanbaatar, Mongolia. (https://commons.wikimedia.org/wiki/File:Diorama,_cavemen_-_National_Museum_of_Mongolian_History.jpg)

were expressed in ritual burial, an example being the Skhul cave, Mount Carmel, Israel, where skeletons painted in red ochre are surrounded by tools (Hovers and Kuhn 2006). Cremation constituted a special way of merging the spirit of the deceased with fire, allowing a passage of the soul to eternity. When civilization rose, burial rituals evolved into grand monuments for the afterlife, represented by the Egyptian pyramids and Chinese imperial burial caves, the latter including entombed entourages intended to serve the ruler in the hereafter, such as Emperor Qin Shi Huang's terracotta army.[9]

Evidence for early small-scale farming is recorded about 23,000 years ago,[10] progressing with the onset of the Neolithic some 12,000 years ago.[11] Now human dreams and illusions could be materialized thanks to the new wealth and resources, including smelting of metal for tools and swords, harnessed to the construction of monuments for the dead and for rituals of death – called war. The premonition of death leads to tension between foresights acquired by the neocortex and the instant reflexes of the mammalian brain (Koestler 1978). Where the cerebral neocortex invents tools and techniques and identifies future dangers, the primitive brain reacts instinctively through defensive/aggressive impulses (Richard Dawkins 1989), with ensuing in internal as well as external inter-personal and inter-tribal conflicts. When equipped with weapons designed by the intelligent brain, from spears to nuclear missiles, such responses lead to destructive violence translated from an individual to the tribal and national scales, and on to the destruction of nature on a regional to global scale. By analogy with infanticide carried out by rival alpha male baboons,[12] running through human history under tribal, racial, national, ideological or religious flags are mass sacrifices of the young, children thrown into Moloch's fire,[13] the Mayan blood cult (Clendinnen 1995) (Fig. 5.3), to women condemning defeated gladiators, to priests promoting crusades and witch hunts, to Easter Island-type ecocide (Fig. 5.4), to the sacrifice of generations as in World War I (Fig. 5.5), to industrial genocide in World War II (Fig. 5.6), to napalm bombing of peasants in their rice fields (Fig. 5.7), appeasing inner demons and angry gods in a quest for immortality.

As in a Maya war song (1300–1521 AD):

There is nothing like death in war,
Nothing like flowering death, so precious to him who gives life,
Far off I see it, my heart yearns for it.

With the onset of space exploration, from the Sputnik to the manned lunar landing and Martian probes, the Galileo space craft and Voyager's interstellar mission, religious mythologies have evolved into a space cult, alluding to colonization of space where presumably *H. sapiens*, raising a blood-stained, tattooed hand toward the stars (Koestler 1978), would proceed to overwhelm the biospheres of new planets. The allusion of the space cult to panspermia and human propagation to other planets and solar systems ignores observations arising from planetary exploration, but barring possible presence of microbes the solar planets appear to be barren. Ideas by proponents of a space race as if life on Earth was seeded by comets appear to be as ideologically motivated as it is lacking in evidence. Comets and space dust

[9] http://www.globalmountainsummit.org/terra-cotta-warriors.html

[10] https://www.sciencedaily.com/releases/2015/07/150722144709.htm

[11] http://www.historyworld.net/wrldhis/PlainTextHistories.asp?ParagraphID=ayh

[12] http://www.pbs.org/wnet/nature/murder-in-the-troop-inside-chacma-baboon-society/2051/

[13] https://carm.org/christianity/miscellaneous-topics/moloch-ancient-pagan-god-child-sacrifice

Fig. 5.3 Mayan blood cult (https://upload.wikimedia.org/wikipedia/commons/0/0e/British_museum173.jpg)

Fig. 5.4 Easter Island statues (https://commons.wikimedia.org/wiki/File:Moai_Rano_raraku.jpg)

Fig. 5.5 (**a**) Gas blinded soldiers, British 55th Division 10_April_1918 (https://de.wikipedia.org/wiki/Gaskrieg_w%C3%A4hrend_des_Ersten_Weltkrieges#/media/File:British_55th_Division_gas_casualties_10_April_1918.jpg). (**b**) Soldier asleep in trench, Thievpal, France, during World War I (https://www.flickr.com/photos/nlscotland/3012796098)

Fig. 5.6 The Auschwitz death camp ovens (https://de.wikipedia.org/wiki/Datei:Verbrennungs%C3%B6fen.JPG)

are not known to host biomolecules. The presence of amino acids is no more evidence for life than, for example, the presence of iron in sea water represent computers.

A new geological era was born on July 16, 1945 with the Trinity test (Taylor 2015), coined here as *Plutocene,* after the element Plutonium whose isotope ^{239}Pu decays to ^{235}U over a half-life of 24,100 years, a time scale consistent with the sequestration of the excess CO_2 emitted during the Late Anthropocene. During the Plutocene the biosphere is dominated by elevated temperatures, analogous to those of the Pliocene

Fig. 5.7 Phan Thị Kim Phúc OOnt (Napalm girl) Viet Nam (https://www.flickr.com/photos/13476480@N07/26169340584)

(2.6–5.3 Ma ago) and the Miocene (~5.3–23 Ma ago), when mean global temperatures were 2–4 °C warmer than pre-industrial temperatures. During the Plutocene acid oceans reach 20–40 m higher than during Pleistocene interglacial periods, inundating deltas and river valleys around the world and with them the sites of ancient and modern civilizations. High temperatures and radioactivity constrain biological activity to radiation-resistant species, mostly the arthropods, allowing proliferation of sophisticated colonial insect habitats. Atmospheric CO_2 residence time on the order of 10^3 to 10^4 years (Solomon et al. 2009; Eby et al. 2009), delay the subsequent glacial cycle from ~20,000 years to perhaps ~50,000 years in the post-Anthropocene (Berger and Loutre 2002) (Fig. 1.21). These factors restrict comparisons of the Plutocene with the Pliocene and Miocene biospheres, in particular since flora and fauna evolved gradually during these periods, as contrasted with the abrupt onset of the Late Anthropocene global heating, driven by an extreme rise in CO_2 of >3 ppm CO_2/year. CO_2 rise between November 1995 (400.16 ppm) and November 1996 (403.53 ppm) reaching 3.37 ppm/year (NOAA 2016[14]) has no known precedents at least as far back as 55 Ma (Glikson 2016).

With estimated profitable carbon reserves in excess of 10,000 billion ton of carbon (GtC) (Fig. 1.19), further emissions could take the atmosphere out of the ice ages back to Mesozoic-like greenhouse conditions, a state during which large parts of the continents were inundated by the sea. Most likely to survive would be the

[14] https://www.esrl.noaa.gov/gmd/ccgg/trends/

grasses and insects, and perhaps some birds, descendants of the fated dinosaurs. A new evolutionary cycle would commence. *Homo sapiens* may survive—the endurance of the species through extreme climate upheavals of the glacial-interglacial periods has equipped humans to withstand the most challenging conditions.

Rapid to extreme warming rates lead to fires consuming large parts of the tropical and sub-tropical forest biosphere (Bowman et al. 2009) and even in Boreal forests in Siberia (Union of Concerned Scientists 2016[15]). While trees, growing over medium to long time frames are prone to fire, grasses with their well protected seeds can survive droughts and fires allowing the expansion of savannah, including some radiation-resistant organisms, mostly arthropods (Koval 1983; Dushimirimana et al. 2012). A resumption of glacial period induced by the Milankovic cycles and a gradual decline in radioactivity would lead to re-emergence of survivors, including burrowing mammals. Depending on the intensity of radioactive contamination, relic hunter-gatherer humans may or may not survive in northern latitudes and in cold high-altitude valleys.

The Sixth mass extinction does not rise exclusively from global warming, and is brought about, separately or in combination, by design or accident, through the reduction of natural habitats (WWF 2016a, b[16]) and the probability of a global nuclear cataclysm. As time goes on and the dice falls, possibilities may become probabilities and probabilities may become certainties, an increasingly likely prospect on a warming planet burdened by energy and resource wars. Following trials on the inhabitants of two Japanese cities, with time, the Damocles sword of Mutual Assured Destruction (MAD) and "*use-them-or-lose them*" strategies (Nuclear Files 1961[17]) can only fall. The hapless inhabitants of planet Earth are given a non-choice between runaway global warming, a transient stadial freeze, and the coup-de-grace of a nuclear winter, followed by further warming due to accumulated greenhouse gases. In the meantime too many members of the species appear to be more concerned with the World Cup than with Planetary survival.

Initially it has not been entirely clear whether the detonation of the first atomic bomb would not result in a chain reaction incinerating the Earth's entire atmosphere, a question on which Edward Teller, "*Father*" of the hydrogen bomb agonized.[18] Further experiments with the Earth are underway. Once the *Large Hadron Collider,*[19] deemed by its designers to be safe,[20] was questioned by Stephen Hawking who stated the unthinkable: '*finding the Higgs-Boson 'God' particle could destroy the universe* (Gillman 2014). The Higgs boson particle is thought to be part of the mechanism that gives matter its mass, but scientists do not fully understand it yet. Pending further science fiction-like experiments dreamt by ethics-free scientists,

[15] http://www.climatehotmap.org/global-warming-locations/western-siberia.html

[16] http://wwf.panda.org/about_our_earth/species/problems/habitat_loss_degradation/

[17] http://www.nuclearfiles.org/menu/key-issues/nuclear-weapons/history/cold-war/strategy/strategy-mutual-assured-destruction.htm

[18] http://mentalfloss.com/article/66160/10-explosive-facts-about-trinity-nuclear-test

[19] http://home.cern/topics/large-hadron-collider

[20] http://www.stfc.ac.uk/research/particle-physics-and-particle-astrophysics/large-hadron-collider/

Fig. 5.8 The Large Hadron Collider based at the European particle physics laboratory CERN, near Geneva in Switzerland (https://apod.nasa.gov/apod/ap080225.html)

Fig. 5.9 Stephen Hawking stated: finding the Higgs-Boson 'God' particle could destroy the universe (https://www.flickr.com/photos/lwpkommunikacio/16203367706)

Earth may not become a black hole. Unfortunately little doubt exists regarding the consequences of the continuing use of the atmosphere, the lungs of the biosphere, as open sewer for carbon gases, or continuing to maintain and expand doomsday nuclear weapons (Figs. 5.8 and 5.9).

By the first decade of the twenty-first century the emission of carbon gases into the atmosphere, totalling over 600 billion tons of carbon (>600 GtC) to date, is generating amplifying feedbacks ensuring a climate calamity. As stated by the renowned oceanographer Wallace Broecker (1987), "*The inhabitants of planet Earth are quietly conducting a gigantic experiment. We play Russian roulette with climate and no one knows what lies in the active chamber of the gun.*"

Where the Nazis constructed gas chambers for millions of victims, runaway climate change and nuclear hair-triggered missile fleets threaten to turn the entire Planet into an open oven on the strength of a *Faustian Bargain*. From the Roman Emperors to the third Reich, the barbarism of empires surpasses that of small marauding tribes. In the name of freedom and human rights they never cease to bomb peasant populations in their rice fields. It is among the wretched of the Earth that true charity is common, where empathy is learnt through suffering.

Planetcide challenges every faith, ideal and social system humans ever held. Individuals are crushed, as in H.G. Wells War of the Worlds (2011),[21] when cells rebelling against the insanity of a murderous global Martian society are destroyed by the parent organism. Planetcide is a child of *Orwellian Newspeak*, (Orwell 1949)[22] where modern societies, underpinned by subterranean drug rings, weapon smuggling networks and intelligence agencies, poison their young's minds with commercial and political lies, a propaganda machine Joseph Goebbels would envy.

Nature is full of examples of viruses destroying their host and animals seducing their prey. Through the use of language some humans have perfected untruths to a high form of art. Defying the scientific method, some succeeded in undermining humanity's desperate attempt at mitigating runaway global warming (Goldblat and Watson 2012) by almost 40 years (Oreskes and Conway 2011).

Having lost the sense of reverence toward the Earth possessed by prehistoric humans, there is no evidence civilization is about to adopt Carl Sagan's sentiment[23]: "*For we are the local embodiment of a Cosmos grown to self-awareness. We have begun to contemplate our origins: star stuff pondering the stars: organized assemblages of ten billion billion billion atoms considering the evolution of atoms; tracing the long journey by which, here at least, consciousness arose. Our loyalties are to the species and the planet. We speak for the Earth. Our obligation to survive is owed not just to ourselves but also to that Cosmos, ancient and vast, from which we spring*" (Carl Sagan 1980).

Cultural evolution occurs at faster rates and potentially out of control rates as compared with biological genetic evolution, culminating with intelligent primates waving instruments of global destruction. As stated by J. Robert Oppenheimer: *the physicists have known sin; and this is a knowledge which they cannot lose,*[24] passing the ultimate weapon to potentially criminal hands. Using a plethora of platitudes humanity is presiding over its own demise and that of much of nature, committing the planet to a greenhouse atmosphere and nuclear chain reactions. Since World War II claimed the lives of about 48 million human beings,[25] the world has become wired to computer-controlled nuclear arms control and command hair-trigger systems capable of triggering a holocaust by error or design, global phone and internet surveillance systems (Van der Sloot 2016), satellite communication networks, video

[21] http://www.shmoop.com/war-of-the-worlds-hg-wells/summary.html

[22] http://orwell.ru/library/novels/1984/english/en_app

[23] https://www.amazon.com/Cosmos-Carl-Sagan/dp/B000055ZOB

[24] https://todayinsci.com/O/Oppenheimer_Robert/OppenheimerRobert-Quotations.htm

[25] http://warchronicle.com/numbers/WWII/deaths.htm

cameras and GPS tracking arrays capable of identifying locations and computer key strokes[26,27,28,29]). A new generation fed by the open-ended untruths of TV commercials broadcast by talking heads, snake oil merchants, celebrities, sport icons, one-liner-tweets and faceless books beaming from fluorescent screens, an ethics-free electronic jungle where illusions and facts are no longer discernible.[30] Plugged to their ear phones sleep-walking they wander aimlessly while their planet is overwhelmed, committing them to an unliveable world.

With the ultimate Faustian Bargain manifested by the "*Price of the Earth*", estimated as "$3000 trillion" according to a "*scientist's planet valuing formulae*",[31] or 5 quadrillion according to another,[32] "*Sapiens*", living in a realm of myths and legends (Koestler 1978) is betraying its birthright. It wakes up for a brief moment from a universal slumber to witness a biosphere dominated by the food chain, a world as cruel as it is beautiful. Possibly an inverse relation may exist between the level of consciousness achieved by a species and its longevity, once it creates machines and processes that it cannot control. If looking into the sun may result in blindness, metaphorically the deep insights into nature that humans have achieved may bear a terrible price.

Human ideologies and "isms" are capable of a reductionist classification into those which promote life and those which perpetrate death. Technological societies dominated by death industries, producing military homicide technologies and fatal fossil fuel emissions, determine their own future. Existential philosophy allows a perspective into, and a way of coping with what defies rational contemplation. Ethical and cultural assumptions of free will rarely govern the behaviour of societies or nations, let alone an entire species. And while the planet may not shed a tear for the demise of technological civilization, momentary *hope* on the individual scale is possible. Going through their black night of the soul members of the species may be rewarded by the emergence of a conscious dignity devoid of illusions, grateful for a fleeting glimpse into the universe for which humans are privileged.

For a species to learn to ignite fire and harness the atom, enhancing its power and energy output and entropy in nature by orders of magnitude greater than its own physical capacity, the species needs to be perfectly wise and responsible, lest its invention gets out of control, engulfing nature. It is unlikely *Homo sapiens* can achieve such levels of wisdom and responsibility.

> *Having pushed a boulder up the mountain all day, turning toward the setting sun, we must consider Sisyphus happy.* (Camus 1942).

---

[26] http://news.bbc.co.uk/2/hi/503224.stm

[27] https://www.jipitec.eu/issues/jipitec-6-3-2015/4318

[28] https://nakedsecurity.sophos.com/2016/04/07/over-75-of-people-lie-on-social-media/

[29] http://www.star2.com/living/viewpoints/2015/11/09/but-then-again-everybody-lies-on-social-media/

[30] http://www.monbiot.com/2017/03/02/screened-out/

[31] http://www.dailymail.co.uk/sciencetech/article-1361145/Earth-worth-3-000-trillion-according-scientists-new-planet-valuing-formula.html

[32] http://www.treehugger.com/natural-sciences/new-formula-values-earth-at-5000000000000000.html

References

Beerling D (2002) CO2 and the end-Triassic mass extinction. Nature 415(6870):386–387
Berger A, Loutre MF (2002) An exceptionally long interglacial ahead? Science 297:1287–1288
Bowman DM et al (2009) Fire in the Earth system. Science 324:481–484
Broecker WS (1987) Unpleasant surprises in the greenhouse? Nature 328:123–126
Camus A (1942) The myth of Sisyphus. Vintage International/Amazon, New York, p 205
Clendinnen I (1995) Aztecs: an interpretation. Cambridge University Press, Cambrdige, 398 pp
Dawkins R (1989) The selfish gene. Oxford University Press, New York, p 351
deMenocal PB (2004) African climate change and faunal evolution during the Pliocene-Pleistocene. Earth Planet Sci Lett 220:3–24
Dushimirimana S et al (2012) Comparison of reproductive traits of regular and irradiated male desert locust Schistocerca gregaria (Orthoptera: Acrididae): evidence of last-male sperm precedence. Biol Open 15(1/3):232–236
Eby N, Zickfeld K, Montenegro A, Archer D (2009) Lifetime of anthropogenic climate change millennial time scales of potential CO_2 and surface temperature perturbations. J Clim 22:2501–2511
Gillman O (2014) Finding the 'God' particle could destroy the universe, warns Stephen Hawking Daily Mail. http://www.dailymail.co.uk/news/article-2746727/Maybe-shouldn-t-looking-quite-hard-God-particle-destroy-universe-warns-Stephen-Hawking.html
Glikson AY (2005) Asteroid/comet impact clusters, flood basalts and mass extinctions significance of isotopic age overlaps. Earth Planet Sci Lett 236:933–937
Glikson AY (2016) Cenozoic mean greenhouse gases and temperature changes with reference to the Anthropocene. Glob Chang Biol 22:3843–3858
Glikson AY, Groves C (2016) Climate, fire and human evolution: the deep time dimensions of the Anthropocene. Springer, Dordrecht, p 226
Goldblat C, Watson AJ (2012) The Runaway Greenhouse: implications for future climate change, geoengineering and planetary atmospheres. Phil Trans Roy Soc 370(1974):4197–4216
Hovers E, Kuhn S (2006) Transitions before the transition: evolution and stability in the middle Paleolithic and middle stone age. Springer, Dordrecht, p 332
Keller G (2005) Impacts volcanism and mass extinction: random coincidence or cause and effect? Aust J Earth Sci 52:725–757
Koestler A (1978) Janus: a summing up. Hutchinson, London, p 384
Kolbert E (2015) The sixth extinction: an unnatural history. Henry Holt and Co, New York, p 307
Koval TM (1983) Intrinsic resistance to the lethal effects of x-irradiation in insect and arachnid cells. Proc Natl Acad Sci 80(15):4752–4755
McElwain JC, Punyasena SW (2007) Mass extinction events and the plant fossil record. Trends Ecol Evol 22(10):548–557
NOAA (2016) Trends in atmospheric carbon dioxide. A global network for measurements of greenhouse gases in the atmosphere. https://www.esrl.noaa.gov/gmd/ccgg/trends/
Nuclear Files 1961. http://nuclearfiles.org/menu/key-issues/nuclear-weapons/issues/accidents/20-mishaps-maybecaused-nuclear-war.htm
Oreskes N, Conway EM (2011) Merchants of doubt: how a handful of scientists obscured the truth on issues from tobacco smoke to global warming. Bloombury Press, New York, p 345
Orwell G (1949) Nineteen eighty-four – the principles of new speak. Penguin Books. http://orwell.ru/library/novels/1984/english/en_app
Sagan C (1980) Cosmos. Random House, New York, p 396
Solomon S, Plattner GK, Knutti R, Friedlingstein P (2009) Irreversible climate change due to carbon dioxide emissions. Proc Natl Acad Sci 106(6):1704–1709
Stanley SM (2016) Estimates of the magnitudes of major marine mass extinctions in earth history. Proc Natl Acad Sci 113:E6325–E6334
Taylor A (2015) 70 years since trinity: the day the nuclear age began. July 16, 2015. https://www.theatlantic.com/photo/2015/07/70-years-since-trinity-when-we-tested-nuclearbombs/398735/

Union of Concerned Scientists (2016) The risk of Nuclear War with China 2016. http://www.ucsusa.org/nuclear-weapons/us-china-relations/risk-nuclear-war-china#.WKzYKbm7oy9
Van der Sloot B (2016) Welcome to the Jungle: the liability of internet intermediaries for privacy violations in Europe. https://www.jipitec.eu/issues/jipitec-6-3-2015/4318
Wells HG (2011) The War of the Worlds. Penquin, London, p 240. http://www.shmoop.com/war-of-the-worlds-hg-wells/summary.html
Wrangham R (2009) Catching fire: how cooking made us human. Basic Books, New York, p 320
WWF (2016a) – How many species are we losing? http://wwf.panda.org/about_our_earth/biodiversity/biodiversity/
WWF (2016b) – Impact of habitat loss on species. http://wwf.panda.org/about_our_earth/species/problems/habitat_loss_degradation/

Chapter 6
Darwinian Evolution, Intelligent Purpose and Mass Extinctions of Species

Abstract As *H. sapiens* is facing its most critical choice (Fig. 6.1), a search for explanations of the 7th mass extinction of species in terms of evolutionary theory and intelligence in nature, while not forthcoming, leads to ultimate questions regarding the origin and evolution of intelligence and the origin of *purpose* in nature.[1]

Like his geologist colleague Charles Lyell[2] (Hallam (1998) who adhering to uniformitarian theory believed in the stability of species, Charles Darwin (1809–1882) rejected the theory that mass extinctions were caused by great catastrophes as promoted by Georges Cuvier (1769–1832).[3] Darwin considered gaps in the geological record to reflect a paucity of observations for limited intervals.[4] Whereas the Cretaceous-Tertiary (K-T) event was well known in Darwin's day, he was convinced the sudden disappearances of species from the fossil record arose from an unrecognized hiatus, an assumption which pertained to the abrupt extermination of families or orders like the Trilobites at the end of the Palaeozoic and of the Dinosaurs and Ammonites at the K-T boundary. There is convincing evidence that at least some of the great mass extinctions of species were triggered by asteroid impacts (Alvarez et al. 1980), volcanic eruptions (Schoene 2009) and methane and hydrogen sulphide emanations (Ward 2008) (Fig. 1.2). Inexplicably the current and 7th mass extinction of species is taking place due to the activity of a living organism – *H. sapiens*.

---

[1] http://progettocosmo.altervista.org/index.php?option=content&task=view&id=87

[2] http://sp.lyellcollection.org/content/143/1/133.full.pdf

[3] http://m.docente.unife.it/giorgio.bertorelle/didattica_insegnamenti/biologia-evoluzionistica-1/StoriaPensieroEvolutivo_1800_Berkeley.pdf

[4] http://www.biology-online.org/articles/role_extinction_evolution/charles_darwin_extinction.html

A.Y. Glikson, *The Plutocene: Blueprints for a Post-Anthropocene Greenhouse Earth*, Modern Approaches in Solid Earth Sciences 13,
DOI 10.1007/978-3-319-57237-6_6

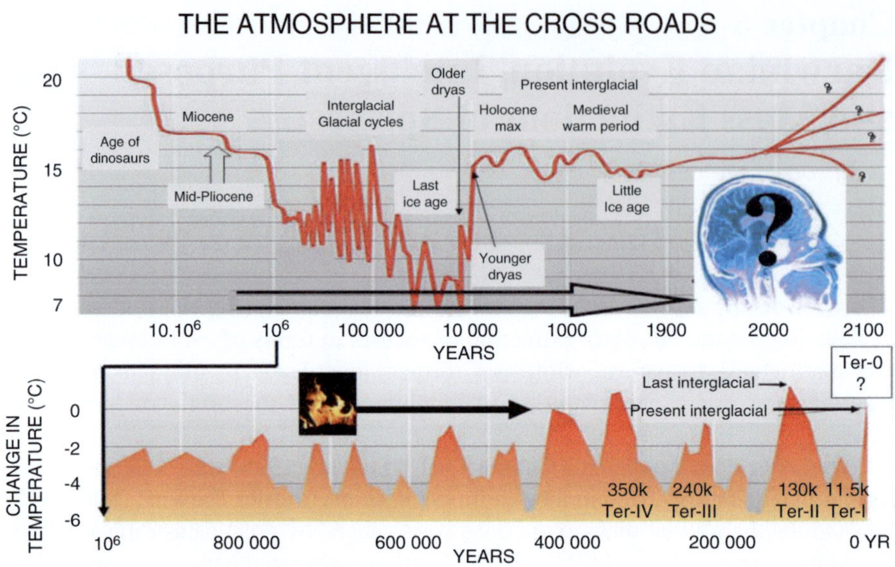

Fig. 6.1 Schematic representation of climate and human evolution: *H. sapiens* at the cross-roads

6.1 Darwinian Evolution

In terms of Darwin's evolutionary theory the phenomenon of a species generating its own demise, and that of much of the biosphere, as *H. sapiens* is in the process of undertaking, remains inexplicable. A possible corollary is Peter Ward's *Medea Hypothesis* (Ward 2008) according to which the Permian-Triassic boundary mass extinction was triggered by hydrogen sulphide emanating from purple-green algae, poisoning the seas and atmosphere at 251 Ma (Ward 2007, 2008), in sharp contrast to James Lovelock's Gaia theory (Lovelock 1995) which espouses homeostatic physical and chemical interactions maintaining Earth's habitability. Manifestly life on Earth evolves in cycles, abruptly terminated either at their inceptions or at an advanced stage, by external events such as asteroid impacts, volcanic eruptions or by an enigmatic self-destruction as the 7th mass extinction of species.

The terrestrial biosphere, constituting a less than 20 km-thick zone constrained by the crustal depth at which bacteria occur and the atmospheric level to which birds can fly, has been repeatedly affected by external forcing, including volcanic events, asteroid and comet impacts, solar insolation, orbital forcing cycles and likely supernovae. Intrinsic biological evolution and diversification through natural selection and adaptation has been repeatedly overprinted, and in some instances almost obliterated, by these events. Both the *Medea hypothesis* (Ward 2008) and the *Cronus hypothesis* (Bradshaw and Brook 2006; Sodhi et al. 2009), hinge on self-destruction. According to These authors "*One can marvel at the incredible diversity of life on Earth, e.g. conservatively > 4 million protists, 16,600 protozoa, 75,000–300,000 helminth parasites, 1.5 million fungi, 320,000 plants, 4–6 million arthropods,*

>6500 amphibians, 10,000 birds and > 5000 mammals) and wonder that there might be something in the 'life makes it easier for life' idea underlying Gaia. However, when one considers that over 99% of all species that have ever existed are today extinct, then a Medea perspective might dominate". The theory however appears to underestimate the role of externally forced environmental factors.

This enigma constitutes one of the fundamental questions hanging over the understanding of natural evolution, such as the origin and fate of the universe, nature of the singularity, the origin of life, the origin and evolution of intelligence (Michio Kaku 1996), the anthropic principle,[5] top-down causation (George Ellis 2005, 2011) and big questions inherent in cosmology (Paul Davies 1984) such as the multiverse (Paul Davies 1984). Inherent in the evolutionary enigma are little-understood top-to-base mechanisms, including the hierarchy of scales and the communication between scales explored among others by George Ellis, who states: "*although the laws of physics explain much of the world around us, we still do not have a realistic description of causality in truly complex hierarchical structures*" (*Physics, complexity and causality*).[6]

6.2 Biological Intelligence

A major question is perceived with regard to the synthesis of the original biomolecules, RNA and DNA, from basic atoms of carbon, oxygen, hydrogen, sulphur and nitrogen, and their evolution into complex forms of evolving and self-repairing life forms, following an apparent trajectory of perceived *purpose,* or the Will to choose. Mutations, natural selection, rearing of the young, survival of the fittest, mass extinctions and other processes inherent in evolution involve a wide range of both accidental events and deliberate decisions and responses to external conditions. Adaptation to external environmental factors, the need to reproduce, and social pressures and coincidences all play a role in evolution. The trajectory of *purpose* intrinsic to cerebral processes in humans and animals, following lines of reasoning and thinking aimed at biological success, are not easily reconciled with accidental factors. On the other hand, according to Richard Dawkins: "*Humans have always wondered about the meaning of life...life has no higher purpose than to perpetuate the survival of DNA…life has no design, no purpose, no evil and no good, nothing but blind pitiless indifference*" (Dawkins 2007).

The *Strong Anthropic Principle,*[7] by claiming an inherent human bias, would appear to deny the objective nature of empirical evidence-based observations and

[5] https://en.wikipedia.org/wiki/Anthropic_principle

[6] http://www.nature.com/nature/journal/v435/n7043/full/435743a.html

[7] The strong anthropic principle (SAP) as explained by John D. Barrow and Frank Tipler states that this is all the case because the universe is in some sense compelled to eventually have conscious and sapient life emerge within it. http://www.physics.sfsu.edu/~lwilliam/sota/anth/SAP_FAP.htm

Fig. 6.2 The Goddess Bastet and The Obsequies of an Egyptian Cat (1886) (https://commons.wikimedia.org/wiki/File:Figurine_of_the_Goddess_Bastet_as_a_Cat_LACMA_AC1992.152.51.jpg, https://commons.wikimedia.org/wiki/File:John_Reinhard_Weguelin_%E2%80%93_The_Obsequies_of_an_Egyptian_Cat_(1886).jpg)

related calculations inherent in the scientific method.[8] A deeply rooted human bias is directed toward the rest of nature, namely the animals where, according to the Seventh Day Creation story, God has given man dominion over the animals, stating "*Let us make mankind in our image, to be like us. Let them be masters over the fish in the ocean, the birds that fly, the livestock, everything that crawls on the earth, and over the earth itself*",[9] sharply contrasted with Pantheistic traditions of reverence of animals, often appointed as gods (Fig. 6.2). This bias is reflected in common albeit non-scientific notions as if animal behaviour hinges purely on genetic, mechanistic, instinctive and innate factors, rather than betraying individual and collective intelligence, as stated[10]: "*Innate behaviours occur in all animals. However, they are less common in species with higher levels of intelligence. Humans are the most intelligent species, and they have very few inate behaviours. The only innate behaviours in humans are reflexes.*" Likewise[11]: "*The phrase "animal intelligence" may introduce a discussion about whether it is meaningful to speak of animals as "intelligent" at all, or whether animal behaviour should instead be thought of as a series of unthinking mechanical responses to stimuli that originate in the animal's internal or external environments, with only humans being capable of conscious thought and flexible responding.*"

Such views, distinguishing between the intelligence and behaviour of species of "high intelligence" and other species may arise from the Anthropic Principle and

[8] As a result advocates of the Anthropic principle believe it is unremarkable that this universe has fundamental constants that happen to fall within the narrow range thought to be compatible with life

[9] https://bibleview.org/en/bible/genesis/7days/

[10] http://www.ck12.org/biology/Innate-Behavior-of-Animals/lesson/Innate-Behavior-of-Animals-BIO/

[11] http://www.assessmentpsychology.com/animal-intelligence.htm

Fig. 6.3 Hermit Crab mounted by sea anemones (https://en.wikipedia.org/wiki/Calliactis_polypus#/media/File:Calliactis_polypus_by_OpenCage.jpg)

are questionable. There can be no doubt regarding the intelligence of numerous species of animals (Gregg 2013), all the way to the Arthropods, as manifested in numerous examples,[12,13]

The purposeful caring behaviour of animals in nurturing their young is a case in point. A snow goose feeding its young, a fish harbouring its infants in its mouth, an octopus or a cockroach sheltering its offspring under its body for years, display little difference from the behaviour of primates and Hominids (WWF 2016a, b[14,15,16]). Watch an ant guiding and supporting a young ant along the trail. Numerous lines of evidence demonstrate collective and individual intelligent cerebral processes in members of the Mammalia, Amphibia and Arthropod classes, the latter recently explored using modern optical-electronic equipment in "*In the Undergrowth*" by David Attenborough,[17] including sophisticated communication, construction, reproduction, parenting, hunting, defence, camouflage, deception and other modes of behaviour requiring human-like intelligence.

Animals use a wide range of physical and recognition methods including electricity (i.e. electric eel), phosphorescence (i.e. glow worms), radar (i.e. bats), star-guided navigation and magnetism (i.e. migratory birds) and solar-scanning navigation (i.e. Cataglyphis ant), betraying human-like genetic and/or learnt sensory and cerebral equipment. Examples of deliberate and likely conscious intelligent cooperation include symbiotic relations between a hermit crab and attached sea anemone,[18] (Nat Geo 2012[19]) (Fig. 6.3) and between Gobi fish and a shrimp[20] (Fig. 6.4).

[12] http://phenomena.nationalgeographic.com/2014/04/22/this-is-how-you-study-the-evolution-of-animal-intelligence/

[13] http://www.assessmentpsychology.com/animal-intelligence.htm

[14] http://www.worldwildlife.org/stories/5-remarkable-animal-moms

[15] https://www.youtube.com/watch?v=e1EPxRuY-eI

[16] http://www.animalplanet.com/wild-animals/1-sea-louse/

[17] http://www.imdb.com/title/tt0760222/

[18] http://animals.mom.me/relationship-between-hermit-crabs-sea-anemones-1857.html

[19] https://www.youtube.com/watch?v=ExV4b77qfww

[20] http://www.liveaquaria.com/PIC/article.cfm?aid=201

Fig. 6.4 Symbiosis between a Gobi fish and a shrimp (Wikipedia Commons) (https://commons.wikimedia.org/wiki/File:Cryptocentrus_and_Alpheus.JPG)

Competition and survival strategies among Mammalia, Amphibians and Arthropods are consistent with those among human, an example being the net-casting by fishermen and the Gladiator spider (Fig. 6.5). Other examples of evidently deliberate intelligent behaviour are birds which pretend to be injured,[21] spiders stealing prey and webs by deception from one another,[22] male insects that remove a competitor's eggs from the female's legs and numerous other examples of intentional intelligent activity.

The architectural design by termite colonies of city-like nests consisting of dedicated nurseries, storage compartments, fungus farming cavities, air conditioning and royal chambers, and the construction of leaf shelters by teams of green ants pulling, knitting and using larvae for application of glue[23] (Figs. 6.6 and 6.7), can hardly be described as innate behaviour but speak of parallels with human-like

[21] https://www.youtube.com/watch?v=xUO9vVkc0FY

[22] https://en.wikipedia.org/wiki/Kleptoparasitism

[23] http://outbackrover.blogspot.com.au/2008/10/green-ants.html

Fig. 6.5 Shared principles of natural intelligence: (a) the Gladiator Spider casting its net (https://commons.wikimedia.org/wiki/File:Flickr_-_ggallice_-_Net-casting_spider.jpg); (b) Fisherman casting his net in a Kiribati lagoon (https://commons.wikimedia.org/wiki/File:Kiribati_fisherman_(10706973774).jpg)

intelligence. The small size of individual insects, analogous to miniature computer chips, cannot be used as a qualitative distinction between insects and humans. There is no escape from an impression these creatures don't operate only on instinct—these creatures can ***think***!

Fig. 6.6 Weaver ants stitching a leaf for a shelter (https://commons.wikimedia.org/wiki/File:AntsStitchingLeave.jpg)

Fig. 6.7 Termites (Macrotermes bellicosus) foraging site with a minor soldier and major workers present. (https://en.wikipedia.org/wiki/Macrotermes_bellicosus#/media/File:Macrotermes_bellicosus_minor_soldier.jpg)

6.3 The Origin of Intelligence

While the evolution of species from the oldest bacteria, cyanobacteria and colonial stromatolites has been progressively documented over the last half century or so (Schopf 2001; Gould 1990), the origin and evolution of biological intelligence, from RNA-DNA biomolecules to the human brain, are shrouded in mystery

(Paul Davies 1999). As distinct from mineral crystals, biomolecules bear genetic codes allowing them to produce new organic substances, proteins, and to diversify, adapt, cross-fertilize and evolve into new forms bearing novel genetic codes.[24] As distinct from inorganic compounds biomolecules and organisms are capable of developing sophisticated *trajectories of* ***purpose***, or ***Will***, including intelligent cooperation, competition, deliberate design, procreative strategies, scientific and artistic endeavours, all the way from bacteria to *H. sapiens* (Bronowski 1973).

No physical laws are known which may account for either the origin or the evolution of biological intelligence, including purposeful intelligent thought and action, corresponding to teleological theory[25]: Genetically inherited innate behaviour and instincts evolved through mutations, environmental effects and natural selection do not appear to be able to answer the question - how did assemblages of carbon, oxygen, hydrogen, nitrogen, and sulphur atoms evolve into intelligent purposeful organisms?

The known laws of physics and theories of mathematical complexity and intelligent design such as by Godel, Turing, Chaitin, Von Newman and others, the religions and faith-based philosophies are unable to provide evidence-based information source. As stated in *Progrettocosmo*[26]: "*According to the biological theory of evolution macroevolution is an unguided, unintelligent, natural process of morphing from a unique common ancestor to all plant and animal species (live or extinct). Darwinists believe that macroevolution is possible thanks to random mutations and natural selection only. Biological organisms are very complex, organized and ordered systems. Complexity, organization and order need information. In an unguided, unintelligent, natural process information doesn't increase per se. Information is not gratis because it needs an intelligent source. Unfortunately Darwinian macroevolution lacks any information source. Those who support Intelligent Design (ID) and creationists claim that origin-of-life and origins-of-species need an information source. The former call this information source the "Designer", the latter call it "God", but anyway ID supporters and creationists share the concept and the need of an information source. They reject Darwinian macroevolution because it doesn't recognize this logical and physical necessity*".

Since Galileo, Copernicus and the publication of Isaac Newton (1642 1726) the *Mathematical Principles of Natural Philosophy*[27] the progressive discovery of the basic laws of nature cannot be assumed to have been completed. New laws can be expected to be discovered, in particular with regard to the origin and evolution of life and intelligence. This may include novel principles and theories of complexity and mathematical information theory[28,29] coding theory and combinatorics may yet account for much of the unknown (Wax 2008). To date to a large extent humans may

[24] http://www.bioinformatics.org/jambw/2/3/TranslationTables.html

[25] http://www.livescience.com/24378-scientists-purpose-nature.html

[26] http://progettocosmo.altervista.org/index.php?option=content&task=view&id=87

[27] http://www.maths.tcd.ie/pub/HistMath/People/Newton/Principia/Bk1Sect1/PrBk1St1.pdf

[28] http://progettocosmo.altervista.org/index.php?option=content&task=view&id=87

[29] http://www.natureworldnews.com/articles/793/20130311/math-model-explains-origin-life.htm

still live in *flatland* (Abbott 1884)[30] unable to recognize new laws and extra-dimensions in nature. Whether unknown laws of natural evolution may provide an explanation for an intelligent species perpetrating the demise of the liveable biosphere, including its own civilization, within the context of Darwinian evolution and Cuvier's catastrophism, remains unknown.

References

Abbott EA (1884) Flatland: a romance of many dimensions. http://www.eldritchpress.org/eaa/FL.HTM

Alvarez LW, Alvarez W, Asaro F, Michel HV (1980) Extra-terrestrial cause for the cretaceous-tertiary extinction: experimental results and theoretical interpretation. Science 208: 1095–1085

Bradshaw CJA, Brook BW (2006) Life and death on Earth – the Cronus hypothesis. J Cosmol 2:201–209

Bronowski J (1973) The ascent of man. Random House, New York, p 352

Davies P (1984) God and the new physics. Simon and Schuster, Inc., New York, p 248

Davies P (1999) The fifth miracle: the search for the origin and meaning of life. Touchstone, New York, p 293

Dawkins R (2007) The God delusion. Pemguin/Random House, London, p 457

Ellis GFR (2005) Physics, complexity and causality. Nature 435:743

Ellis GFR (2011) Top-down causation and emergence: some comments on mechanisms. Interface Focus 2:126–140, The Royal Society. 29 September 2011

Gould J (1990) Wonderful life: the burgess shale and the nature of history. Angus and Robertson, Sydney, p 347

Gregg J (2013) A new frontier in animal intelligence. Scient Am 309(2)

Hallam A (1998) Lyell's views on organic progression, evolution and extinction. Geol Soc Lond Spec Publ 143:133–136

Kaku M (1996) The origin of intelligence, Big think. http://bigthink.com/big-think-tv/the-origin-of-intelligence

Lovelock J (1995) Gaia: a new look at life on Earth. Oxford University Press, Oxford, p 147

Schoene B (2009) Correlating the end-Triassic mass extinction and flood basalt volcanism at the 100 ka level. Geology 38:387–390

Schopf JW (2001) Cradle of life: the discovery of Earth's earliest fossils. Princeton University Press, Princeton, p 361

Sodhi NS, Brook BW, Bradshaw JA (2009) Causes and consequences of species extinctions. Princeton guide to Ecology, p 514. http://connection.ebscohost.com/c/book-chapters/53752656/chapter-v-1-causes-consequences-species-extinctions

Ward PD (2007) Under a green sky: global warming, the mass extinctions of the past, and what they can tell us about our future. Hareer and Collins, New York, p 205

Ward P (2008) A theory of Earth's Mass extinctions (TED2008 19:41 Filmed Feb 2008). https://www.ted.com/talks/peter_ward_on_mass_extinctions

Wax H (2008) Using math to explain how life on earth began. Sci Am October 2008. https://www.scientificamerican.com/article/can-math-solve-origin-of-life/

WWF (2016a) – How many species are we losing? http://wwf.panda.org/about_our_earth/biodiversity/biodiversity/

WWF (2016b) – Impact of habitat loss on species. http://wwf.panda.org/about_our_earth/species/problems/habitat_loss_degradation/

[30] http://www.eldritchpress.org/eaa/FL.HTM

Epilogue

Given the close comparisons between human intelligence and that of animals (Sects. 6.2 and 6.3), since about 2 million years ago and in particular since the early Neolithic[1] about 10,200 years ago *H. sapiens* has developed novel techniques unrivalled by any species, including the key discovery of the mastery of fire (Wrangham 2009). The decoding of the basic laws of physics, discovery of electricity, magnetism and nuclear power, and research spanning the range from subatomic particles to astronomy, has given rise to an enigma best captured in Carl Sagan's words (Cosmos 1980):

> *For we are the local embodiment of a Cosmos grown to self-awareness. We have begun to contemplate our origins: star-stuff pondering the stars; organised assemblages of ten billion-billion-billion atoms considering the evolution of atoms; tracing the long journey by which, here at least, consciousness arose. Our loyalties are to the species and the planet. WE speak for the Earth. Our obligation to survive is owed not just to ourselves but also to that Cosmos, ancient and vast, from which we spring.*

That a species which discovered the very natural laws which govern its own cerebral processes, allowing it to gain insights into the principles of physics and the Universe, would perpetrate its own demise and that of much of nature, defies comprehension, inviting poetic expression: Perhaps, by analogy to the eternal conflict between positive and negative subatomic particles inherent in the Universe, and between life and death as symbolized by the Hindu mythological Brahma-Vishnu-Shiva cycle, human ideals cannot survive. Much as looking into the sun for long periods results in blindness, deciphering the secret codes of the universe may bear a terrible price.

[1] https://www.britannica.com/event/Neolithic-Period

A.Y. Glikson, *The Plutocene: Blueprints for a Post-Anthropocene Greenhouse Earth*, Modern Approaches in Solid Earth Sciences 13,
DOI 10.1007/978-3-319-57237-6

The Universal Web

Tiny wispy strings
Cryptic little things
Shape universes, sing
Oh take me on your wings.

X-apparitions chase
Light speed - a cosmic race
Appears from quantum space
Oblivious to my race.

DNA chains extend
RNA healers bend
Repair a link, append
life's wonder: self-amend.

RNA chains collapse
A lover's final gasp
Queen Cleo's sacrifice
While gods are playing dice.

Unseen hands strum guitar
Bells' echoes reach afar
No strings attached, no bar
Can bind a young blue star.

My thought waves rise and ebb
Brains weave a spider's web
Probe truths or make believe
Loft crests I can't achieve.

Web sites each other find
A net's collective mind
Tripwire humankind
A cyclops stumbling blind.

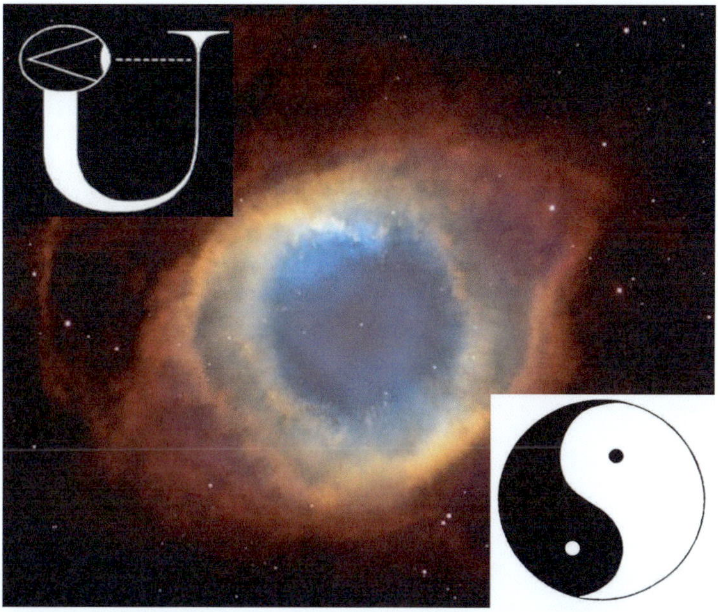

Figure Epilogue 1 The Ring Nebula – a metaphoric eye of the Universe (http://maxpixel.freegreat-picture.com/Ngc-7293-Planetary-Fog-Helix-Nebula-11155). Insets: (**a**) the U symbol, designed by John Wheeler (http://www.physicsoftheuniverse.com/scientists_wheeler.html), symbolizing human perception of the Big Bang; (**b**) the yin and Yen metaphor

About the Author

Andrew Yoram Glikson, An Earth and paleo-climate scientist, studied geology at the University of Jerusalem and graduated at the University of Western Australia in 1968. He conducted geological surveys of ancient geological formations in central and western Australia, Queensland, South Africa, India and Canada; studied large asteroid impacts, including effects on the atmosphere and oceans and mass extinctions of species. Since 2005 he has been studying the relations between climate, fire and human evolution. He was active in communicating nuclear issues and climate evidence to the public and parliament through papers, lectures, conferences and presentations.

A.Y. Glikson, *The Plutocene: Blueprints for a Post-Anthropocene Greenhouse Earth*, Modern Approaches in Solid Earth Sciences 13,
DOI 10.1007/978-3-319-57237-6

References

Aarkrog A (1998) A retrospect of anthropogenic radioactivity in the global marine environment. Radiat Prot Dosimetry 75(1–4):23–31. Oxford Academic Press, Oxford
Aarkrog A (2003) Input of anthropogenic radionuclides into the World Ocean. Deep-Sea Res II Top Stud Oceanogr 50:2597–2606
Abbott EA (1884) Flatland: a romance of many dimensions. http://www.eldritchpress.org/eaa/FL.HTM
Abbott AN et al (2015) Constraints on ocean circulation at the Paleocene–Eocene thermal maximum from neodymium isotopes. Clim Past 12:837–847
Adler J (2013) Why fire makes us human. Smithsonian.com. http://www.smithsonianmag.com/science-nature/why-fire-makes-us-human-72989884/
Alvarez LW, Alvarez W, Asaro F, Michel HV (1980) Extra-terrestrial cause for the cretaceous-tertiary extinction: experimental results and theoretical interpretation. Science 208:1095–1085
Barber DC et al (1999) Forcing of the cold event of 8,200 years ago by catastrophic drainage of Laurentide lakes. Nature 400:344–348
Barnosky AD (2015) Approaching a state shift in Earth's biosphere. Nature 486:52–58
Bauer S (2005) Radiation exposure due to local fallout from Soviet atmospheric nuclear weapons testing in Kazakhstan: solid cancer mortality in the Semipalatinsk historical cohort 1960–1999. Radiat Res 164:409–419
Beerling D (2002) CO2 and the end-Triassic mass extinction. Nature 415(6870):386–387
Beerling DJ, Berner RA (2002) Biogeochemical constraints on the Triassic-Jurassic boundary carbon cycle event. Glob Biogeochem Cycles 16(3):1036
Beerling DJ, Lomax BH, Royer DL, Upchurch GR, Kump LR (2002) An atmospheric pCO2 reconstruction across the cretaceous-tertiary boundary from leaf mega fossils. Proc Natl Acad Sci U S A 99(12):7836–7840
Beerling DJ, Royer D (2011) Convergent Cenozoic CO 2 history. Nat Geosci 4:418–420
Berger A, Loutre MF (2002) An exceptionally long interglacial ahead? Science 297:1287–1288
Berner RA (2004) A model for calcium, magnesium and sulfate in seawater over Phanerozoic time. Am J Sci 304:438–453
Berner RA (2006) GEOCARBSULF: a combined model for Phanerozoic atmospheric O2 and CO2. Geochim Cosmochim Acta 70:5653–5664
Berner RA, Vanderbrook JM, Ward PD (2007) Oxygen and evolution. Science 316:557–558

A.Y. Glikson, *The Plutocene: Blueprints for a Post-Anthropocene Greenhouse Earth*, Modern Approaches in Solid Earth Sciences 13,
DOI 10.1007/978-3-319-57237-6

Bonabeau E et al (1999) Swarm intelligence: from natural to artificial systems. Oxford University Press, Santa Fe Institute Studies in the Sciences of Complexity, JASSS. http://jasss.soc.surrey.ac.uk/4/1/reviews/kluegl.html
Bordner AS (2016) Measurement of background gamma radiation in the northern Marshall Islands. Proc Natl Acad Sci 113(25):6833–6838
Bowen GJ et al (2015) Two massive, rapid releases of carbon during the onset of the Paleocene–Eocene thermal maximum. Nat Geosci 8:44–47
Bowman DM et al (2009) Fire in the Earth system. Science 324:481–484
Bradshaw CJA, Brook BW (2006) Life and death on Earth – the Cronus hypothesis. J Cosmol 2:201–209
Braun H, Ditlevsen P, Chialvo DR (2008) Solar forced Dansgaard-Oeschger events and their phase relation with solar proxies. Geophys Res Lett 35(6):L06703–L06802
Broecker WS (1987) Unpleasant surprises in the greenhouse? Nature 328:123–126
Broecker WS (2000) Abrupt climate change: causal constraints provided by the paleoclimate record. Earth Sci Rev 51:137–154
Broecker WS (2006) Abrupt climate change revisited. Glob Planet Chang 54:211–215
Broecker WS, Stocker TF (2006) The Holocene CO2 rise: anthropogenic or natural? Eos 87(3):27
Bronowski J (1973) The ascent of man. Random House, New York, p 352
Buesseler KO (1997) The isotopic signature of fallout in the North Pacific. J Environ Radioact 36:69–83
Bulletin of the Atomic Scientists (2017) vol 73, Issue 1. http://thebulletin.org/2017/january
Camus A (1942) The myth of Sisyphus. Vintage International/Amazon, New York, p 205
Carlson AE (2010) What caused the younger dryas cold event? Geology 38:383–384
CDIAC (Carbon Dioxide Information Analysis Centre) (2017) 800,000-year Ice-Core records of atmospheric carbon dioxide (CO2). US Department of Energy, Oak Ridge http://cdiac.ornl.gov/trends/co2/ice_core_co2.html
Ceballos G et al (2015) Accelerated modern human–induced species losses: entering the sixth mass extinction. Sci Adv 1(5):e1400253
Chandler J (2011) Feeling the heat. Melbourne University Press, Melbourne, p 292
Chandler M et al (1994) Joint investigations of the middle Pliocene climate II: GISS GCM Northern Hemisphere results. Glob Planet Chang 9:197–219
Clendinnen I (1995) Aztecs: an interpretation. Cambridge University Press, Cambrdige, 398 pp
Cortese G, Abelmann A, Gersonde A (2007) The last five glacial-interglacial transitions: a high resolution 450,000-year record from the subantarctic Atlantic. Paléo 22:PA4203
Crutzen PJ, Birks JW (1982) The atmosphere after a nuclear war: twilight at noon. Ambio 11(2/3):114–125
CTBTO (Comprehensive Nuclear-Test-Ban Treaty) (2012) https://www.ctbto.org/specials/who-we-are/
Cui Y, Kump LR, Ridgwell AJ, Charles AJ, Junium CK, Diefendorf AF, Freeman KH, Urban NM, Harding IC (2011) Slow release of fossil carbon during the Palaeocene–Eocene thermal maximum. Nat Geosci 4:481–485
Cutter GA et al (1979) Deposition and accumulation of plutonium isotopes in Antarctica. Nature 279:628–629
Davies P (1984) God and the new physics. Simon and Schuster, Inc., New York, p 248
Davies P (1999) The fifth miracle: the search for the origin and meaning of life. Touchstone, New York, p 293
Dawkins R (1989) The selfish gene. Oxford University Press, New York, p 351
Dawkins R (2007) The God delusion. Pemguin/Random House, London, p 457
De Chazal J, Rounsevell MDA (2009) Land-use and climate change within assessments of biodiversity change: a review. Glob Environ Chang 19:306–315
deMenocal PB (2004) African climate change and faunal evolution during the Pliocene-Pleistocene. Earth Planet Sci Lett 220:3–24

Deryabina TG et al (2015) Long-term census data reveal abundant wildlife populations at Chernobyl. Curr Biol 25(19):R824–R826

Dong W, Zheng J, Guo Q, Yamada M, Pan S (2010) Characterization of plutonium in deep-sea sediments of the Sulu and South China Seas. J Environ Radioact 8:622–629

Dushimirimana S et al (2012) Comparison of reproductive traits of regular and irradiated male desert locust Schistocerca gregaria (Orthoptera: Acrididae): evidence of last-male sperm precedence. Biol Open 15(1/3):232–236

Eby N, Zickfeld K, Montenegro A, Archer D (2009) Lifetime of anthropogenic climate change millennial time scales of potential CO 2 and surface temperature perturbations. J Clim 22:2501–2511

Ellis GFR (2005) Physics, complexity and causality. Nature 435:743

Ellis GFR (2011) Top-down causation and emergence: some comments on mechanisms. Interface Focus 2:126–140, The Royal Society. 29 September 2011

Espanol M et al (2016) Spatial and temporal Antarctic Ice Sheet mass trends, glacio-isostatic adjustment, and surface processes from a joint inversion of satellite altimeter, gravity, and GPS data. J Geophys Res 121:182–200

Evans G, Ogilvie-White T, Ramesh T (2015) Nuclear weapons the state of play 2015. Australian National University, Canberra

First evidence of farming in Mideast 23,000 years ago (2015) Sci News. https://www.sciencedaily.com/releases/2015/07/150722144709.htm

Foster GI, Rohling EJ (2013) Relationship between sea level and climate forcing by CO2 on geological timescales. Proc Natl Acad Sci 110:1209–1214

France-Lanford C, Derry LA (1997) Organic carbon burial forcing of the carbon cycle from Himalayan erosion. Nature 390:65–67

Ganopolski A, Rahmstorf S (2002) Abrupt glacial climate changes due to stochastic resonance. Phys Rev Lett 88:3–6

Ganopolski A, Winkelmann R, Schellnhuber HJ (2016) Critical insolation–CO2 relation for diagnosing past and future glacial inception. Nature 529:200–203

Garzione CN (2008) Surface uplift of Tibet and Cenozoic global cooling. Geology 36:1003–1004

Gillman O (2014) Finding the 'God' particle could destroy the universe, warns Stephen Hawking Daily Mail. http://www.dailymail.co.uk/news/article-2746727/Maybe-shouldn-t-lookingquite-hard-God-particle-destroy-universe-warns-Stephen-Hawking.html

Glasstone S, Dolan PJ (1950/1977) The effects of nuclear weapons. US Dept Defense and Energy Research and Development Administration. http://www.deepspace.ucsb.edu/wp-content/uploads/2013/01/Effects-of-Nuclear-Weapons-1977-3rd-edition-complete.pdf

Glikson AY (2005) Asteroid/comet impact clusters, flood basalts and mass extinctions significance of isotopic age overlaps. Earth Planet Sci Lett 236:933–937

Glikson AY (2013a) Fire and human evolution: the deep-time blueprints of the Anthropocene. Anthropocene 3:89–92

Glikson AY (2013b) The asteroid impact connection of planetary evolution with special reference to large Precambrian and Australian impacts. Springer, Dordrecht, p 149

Glikson AY (2014) Evolution of the atmosphere, fire and the Anthropocene climate event horizon. Springer, Dordrecht, p 174

Glikson AY (2016) Cenozoic mean greenhouse gases and temperature changes with reference to the Anthropocene. Glob Chang Biol 22:3843–3858

Glikson AY, Groves C (2016) Climate, fire and human evolution: the deep time dimensions of the Anthropocene. Springer, Dordrecht, p 226

Global Carbon Project (2016) Global Methane Budget. http://www.globalcarbonproject.org/methanebudget/

Global Nuclear Arsenals (2009) Approximately 23,335 weapons with ~ 6400 MT (megatons) yield. http://www.nucleardarkness.org/include/nucleardarkness/files/global_nuclear_arsenal_in_number_2009.pdf

Goldblat C, Watson AJ (2012) The Runaway Greenhouse: implications for future climate change, geoengineering and planetary atmospheres. Phil Trans Roy Soc 370(1974):4197–4216
Gould J (1990) Wonderful life: the burgess shale and the nature of history. Angus and Robertson, Sydney, p 347
Gradstein FM, Ogg JG (2004) Geological time scale – why, how and where next. Taylor Fr Lethaia 37:175–181
Gregg J (2013) A new frontier in animal intelligence. Scient Am 309(2)
Grey K, Walter R, Calver CR (2003) Neoproterozoic biotic diversification: snowball Earth or aftermath of the Acraman impact? Geology 31:459–462
Guosheng Y et al (2015) Plutonium concentration and isotopic ratio in soil samples from central-eastern Japan collected around the 1970s. Scientific reports US National Library of Medicine, National Institutes of Health. Sci Rep 5:9636
Hallam A (1998) Lyell's views on organic progression, evolution and extinction. Geol Soc Lond Spec Publ 143:133–136
Hallinan C (2016) We may be at a greater risk of nuclear catastrophe than during the cold war. Foreign Policy In Focus. http://fpif.org/may-greater-risk-nuclear-catastrophe-cold-war/
Hamilton C (2010) A Requiem for a Species. Allen and Unwin, Sydney, p 286
Hancock JJ (2014) The release and persistence of radioactive anthropogenic nuclides. Geol Soc Lond 395:265–281
Hansen JE (2007) Scientific reticence and sea level rise. Environ Res Lett 2:024002
Hansen JE et al (2007) Climate change and trace gases. Phil Trans R Soc A 365A:1925–1954
Hansen JE et al (2008) Target atmospheric CO2: where should humanity aim? Open Atmos Sci J 2:217–231
Hansen JE (2010) Storms of my grandchildren: the truth about the coming climate catastrophe and our last chance to save humanity. Bloomsbury, New York, p 303
Hansen JE (2012) Climatologist James Hansen on "Cowards in Our Democracies". Think Progress https://thinkprogress.org/climatologist-james-hansen-on-cowards-in-our-democracies-c197139f2df9#.1wxe1ict8
Hansen JE, Sato M (2012) Paleoclimate implications for human-made climate change. Clim Chang 2012:21–47
Hansen JE et al (2011) Earth's energy imbalance and implications. Atmos Chem Phys 11:13421–13449. doi:10.5194/acp-11-13421-2011
Hansen JE et al (2012) Perception of climate change. http://www.pnas.org/content/109/37/E2415.abstract
Hansen JE et al (2016) Ice melt, sea level rise and superstorms: evidence from paleoclimate data, climate modeling, and modern observations that 2C global warming could be dangerous. Atmos Chem Phys 16:3761–3812
Hardy EP et al (1973) Global inventory and Distribution of fallout plutonium. Nature 241:444–445
Hatch H et al (2005) The Chernobyl disaster: cancer following the accident at the Chernobyl nuclear plant. Epidemiol Rev 27(1):56–66
Hawking Steven Finding the God' particle could destroy the universe, warns Stephen Hawking. Ollie Gillman for Mail Online. http://www.dailymail.co.uk/news/article-2746727/Maybe-shouldn-t-looking-quite-hard-God-particle-destroy-universe-warns-Stephen-Hawking.html
Haywood AM et al (2007) A permanent El Niño–like state during the Pliocene? Paleoceanography 22(1)
Hoegh-Guldberg O (2007) Coral Reefs under rapid climate change and ocean acidification. Science 5857:1737–1742
Hovers E, Kuhn S (2006) Transitions before the transition: evolution and stability in the middle Paleolithic and middle stone age. Springer, Dordrecht, p 332
Hudson SR (2011) Estimating the global radiative impact of the sea-ice-albedo feedback in the Arctic. J Geophys Res 116:D16102

Hyperphysics (2016) Department of Physics and Astronomy, Georgia State University. http://hyperphysics.phy-astr.gsu.edu/hbase/NucEne/bomb.html
IEER (Institute for Energy and Environmental Research) (1991) Radioactive heaven and Earth. http://ieer.org/wp/wp-content/uploads/1991/06/RadioactiveHeavenEarth1991.pdf
Ikle FC (2015) Worldwide Effects of Nuclear War. http://www.atomicarchive.com/Docs/Effects/wenw_foreword.shtml
INEA (International Nuclear Energy Agency) Nuclear Safety & Security. The International Nuclear and Radiological Event Scale. http://www-ns.iaea.org/tech-areas/emergency/ines.asp
International Society for Reef Studies Climate Change Threatens the Survival of Coral Reefs. http://coralreefs.org/wp-content/uploads/2014/03/ISRS-Consensus-Statement-on-Coral--Bleaching-Climate-Change-FINAL-14Oct2015-HR.pdf
IPCC (2007a) Climate change 2007: working Group I: the physical science basis. https://www.ipcc.ch/publications_and_data/ar4/wg1/en/ch6s6-3-2.html
IPCC (2007b) Climate change 2007: working Group II: impacts, adaptation and vulnerability. https://www.ipcc.ch/publications_and_data/ar4/wg2/en/ch4s4-4-10.html
IPCC (2016) Working group I: the scientific basis. http://www.ipcc.ch/index.htm
Ivanov A (1996) 1950's Soviet A-bomb tests still claim victims. Albion Monitor/News. http://www.albionmonitor.com/9611a/russia1950nuke.html
Johnston R (2008) Multi-megaton tests: the largest Nuclear tests. http://www.johnstonsarchive.net/nuclear/tests/multimegtests.html
Josefsson D (1998) Anthropogenic radionuclides in the Arctic Ocean – Distribution and pathways. Lund University. http://www.iaea.org/inis/collection/NCLCollectionStore/_Public/29/039/29039179.pdf
Kaku M (1996) The origin of intelligence, Big think. http://bigthink.com/big-think-tv/the-origin-of-intelligence
Kasting JF, Donahue TM (1980) The evolution of atmospheric ozone. J Geophys Res 85:3255–3263
Keller G (2005) Impacts volcanism and mass extinction: random coincidence or cause and effect? Aust J Earth Sci 52:725–757
Keller G et al. (2012) Volcanism, impacts and mass extinctions. Geoscientist, The Geological Society. https://www.geolsoc.org.uk/Geoscientist/Archive/November-2012/Volcanism-impacts-and-mass-extinctions-2
Klein RG, Edgar B (2002) The Dawn of human culture. Wiley, New York, p 288
Koestler A (1978) Janus: a summing up. Hutchinson, London, p 384
Koide H (2013) Japan Professor: I believe airborne release of cesium-137 from Fukushima equals 400 to 500 Hiroshima nuclear bombs — another 400 to 500 bombs worth has already flowed into Pacific Ocean. ENE news, march 13th, 2013
Kolbert E (2007) Field notes from a catastrophe: man, nature, and climate change. Bloomsbury, London, p 279
Kolbert E (2015) The sixth extinction: an unnatural history. Henry Holt and Co, New York, p 307
Kompanichenko V (2000) Average lifetime of an intelligent civilization estimated on its global cycle. SAO/NASA Astrophysics Data System, 213, p 438. http://adsabs.harvard.edu/full/2000ASPC..213..437K
Koval TM (1983) Intrinsic resistance to the lethal effects of x-irradiation in insect and arachnid cells. Proc Natl Acad Sci 80(15):4752–4755
Kregenow DA, Swenson ER (2002) The lung and carbon dioxide: implications for permissive and therapeutic hypercapnia. Eur Respir J 20:6–11, http://erj.ersjournals.com/content/20/1/6
Kump LP (2011) The last great global warming. Scientific American, July 2011, 58–61. https://www.scientificamerican.com/article/the-last-great-global-warming/
Kurschner WM et al (2007) The impact of Miocene atmospheric carbon dioxide fluctuations on climate and the evolution of terrestrial ecosystems. Proc Natl Acad Sci 105(2):449–453

Kutzbach JE et al (2010) Climate model simulation of anthropogenic influence on greenhouse-induced climate change (early agriculture to modern): the role of ocean feedbacks. Clim Chang 99:351–381

Lambeck K et al (2014) Sea level and global ice volumes from the last glacial maximum to the Holocene. Proc Natl Acad Sci 111:15296–15303

Lamont Doherty Observatory M (2016) Ocean circulation implicated in past abrupt climate changes. https://www.sciencedaily.com/releases/2016/06/160630145000.htm

Le Quere C et al (2016) Global carbon budget 2016. Earth Syst Sci Data 8:605–649

Lee J (2013) Ten years of Fukushima radiation crossing the Pacific Ocean. CVnews 22 October. https://climateviewer.com/2013/10/22/ten-years-of-fukushima-radiation-crossing-the-pacific-ocean/#prettyPhoto

Leitner K (2004) WHO's involvement in the Chernobyl accident: past and present. http://www.who.int/ionizing_radiation/chernobyl/Overview_WHO_past_involvement.pdf

Lenton TM, Held H, Kriegler E, Hall JW, Lucht W, Rahmstorf S, Schellnhuber HJ (2009) Tipping points in the Earth system. Proc Natl Acad Sci U S A 105:1786–1793

Lewis AR et al (2007) Major middle Miocene climate change and the extinction of tundra communities: evidence from the transantarctic Mountains. US Geol Surv Extended Abstract 135

Lomax ME (2013) Biological consequences of radiation-induced DNA damage: relevance to radiotherapy. Clin Oncol 25(10):578–585

Lovelock J (1995) Gaia: a new look at life on Earth. Oxford University Press, Oxford, p 147

Luisa C (2011) Fukushima radioactive cesium leaks 'equal 168 Hiroshimas. Natural News. http://www.naturalnews.com/033460_Fukushima_Hiroshima.html

Lunt DJ et al (2008) Closure of the Panama Seaway during the Pliocene: implications for climate and Northern Hemisphere glaciation. Clim Dyn 30:1–18

Lynas M (2008) Six degrees: our future on a hotter planet. Harper-Collins e-book, New York

Marais EN (1937/2009) The soul of the white ant. Osiran Books, p 168

Martin B (1988) Nuclear winter: science and politics. Science and Public Policy, Vol. 15, 321–334. http://www.bmartin.cc/pubs/88spp.html

Martin D (2016) Risk of nuclear attack rises. http://www.cbsnews.com/news/60-minutes-risk-of-nuclear-attack-rises/

McElwain JC, Punyasena SW (2007) Mass extinction events and the plant fossil record. Trends Ecol Evol 22(10):548–557

McVicker C (1954) ICE case studies, Novaya Zemlya. http://www1.american.edu/ted/ice/novalya.htm

Meralli Z (2009) Did China's nuclear tests kill thousands and doom future generations? Sci Am. July 2009. https://www.scientificamerican.com/article/did-chinas-nuclear-tests/

Messenger S New formula values Earth at $5,000,000,000,000,000. Tree hugger. http://www.treehugger.com/natural-sciences/new-formula-values-earth-at-5000000000000000.html

Midgley M (1984) Biological and cultural evolution. The institute for cultural research, monograph series no. 20, London NW2 3BW. http://www.i-c-r.org.uk/publications/monographarchive/Monograph20.pdf

Miklós V (2013) The tragic story of the Semipalatinsk Nuclear Test Site. http://io9.gizmodo.com/5988266/the-tragic-story-of-the-semipalatinsk-nuclear-test-site

MIT Technology Review (2012) How likely is a runaway Greenhouse Effect on Earth? https://www.technologyreview.com/s/426608/how-likely-is-a-runaway-greenhouse-effect-on-earth/

Moloch, the ancient pagan God of child sacrifice. https://carm.org/christianity/miscellaneous-topics/moloch-ancient-pagan-god-child-sacrifice

Morelle R (2016) Caribbean coral reefs' could vanish in 20 years. BBC News. http://www.bbc.com/news/science-environment-28113331

Mullen L (2004) Multiple Impacts. Astrobiology Magazine. http://www.astrobio.net/news-exclusive/multiple-impacts/

NASA/PenState (1991) Ancient Climate Events: Paleocene Eocene Thermal Maximum. https://www.e-education.psu.edu/earth103/node/639

NASA (2013) Amplified greenhouse effect shifts north's growing season. http://www.nasa.gov/home/hqnews/2013/mar/HQ_13-069_Northern_Growing_Seasons.html

NASA (2016) The Sunspot Cycle. https://solarscience.msfc.nasa.gov/SunspotCycle.shtml

National Academies Press (2017) National Academy of Sciences. Linkages between arctic warming and mid-latitude weather patterns, p 38. https://www.nap.edu/read/18727/chapter/2

NEI (Nuclear Energy Institute) (2015) Fact sheets. https:// www.nei.org/master-document-folder/backgrounders/fact-sheets/chernobyl-accident-and-its-consequences

NOAA (2016) Trends in atmospheric carbon dioxide. A global network for measurements of greenhouse gases in the atmosphere. https://www.esrl.noaa.gov/gmd/ccgg/trends/

NPR Books (2014) Nuclear 'Command And Control': a history of false alarms and near catastrophes. http://www.npr.org/2014/08/11/339131421/nuclear-command-and-control-a-history-of-false-alarms-and-near-catastrophes

Olivier S et al (2004) Plutonium from global fallout recorded in an ice Core from the Belukha glacier, Siberian Altai. Environ Sci Technol 38(24):6507–6512

Olivier JGJ et al (2014) Trends in global CO2 emissions. PBL Netherlands Environmental Assessment Agency, The Hague, p 59

Oreskes N, Conway EM (2011) Merchants of doubt: how a handful of scientists obscured the truth on issues from tobacco smoke to global warming. Bloombury Press, New York, p 345

Orwell G (1949) Nineteen eighty-four – the principles of new speak. Penguin Books. http://orwell.ru/library/novels/1984/english/en_app

Pagani M et al (2006) Arctic hydrology during global warming at the Palaeocene/Eocene thermal maximum. Nature 442:671–675

Pearce F (2015) Shocking state of world's riskiest nuclear waste site. New Scientist. 21 January 2015.https://www.newscientist.com/article/mg22530053-800-shocking-state-of-worlds-riskiest-nuclear-waste-site/

Pen State Sea level change during the last 5 million years – Exploring the Pliocene-Pleistocene interval. https://www.e-education.psu.edu/earth107/node/901

Petit JR et al (1999) 420,000 years of climate and atmospheric history revealed by the Vostok deep Antarctic ice core. Nature 399:429–436

Phillips AF (1998/2017) 20 mishaps that might have started accidental nuclear war. Nuclear FilesOrg.http://nuclearfiles.org/menu/key-issues/nuclear-weapons/issues/accidents/20-mishaps-maybe-caused-nuclear-war.htm

Ploughshares (2016) World Nuclear Weapon Stockpile. http://www.ploughshares.org/world-nuclear-stockpile-report

Rabi T (2012) The nuclear disaster of Kyshtym 1957 and the politics of the cold war. Arcadia – Environment and Society Portal. http://www.environmentandsociety.org/arcadia/nuclear-disaster-kyshtym-1957-and-politics-cold-war

Radiation effects of a nuclear bomb https://www3.nd.edu/~nsl/Lectures/phys205/pdf/Nuclear_Warfare_9.pdf

Radiation Information Network – Radioactivity in Nature http://www.physics.isu.edu/radinf/natural.htm

RADNET (2016) (since 1996) Information about source points of anthropogenic radioactivity. An information resource for persons interested in the public safety consequences and radio-ecological impact of nuclear accidents and industries. http://www.davistownmuseum.org/cbm/Rad14.html

Rahmstorf S (2003) Timing of abrupt climate change: a precise clock. Geophys Res Lett 30(10):17.1–17.4

Rahmstorf SR, Coumou D (2011) Increase of extreme events in a warming world. Proc Natl Acad Sci U S A 108:17905–17909

Rahmstorf S et al (2015) Exceptional twentieth-century slowdown in Atlantic Ocean overturning circulation. Nat Clim Chang 5:475–480

Raymo ME (2009) PLIOMAX: Pliocene maximum sea level project. PAGES News 17(2). http://moraymo.us/wp-content/uploads/2014/04/2009_raymoetal.pdf

Raymo ME, Ruddiman WF (1992) Tectonic forcing of late Cenozoic climate. Nature 359:117–122

Raymo MR et al (2011) Departures from eustasy in Pliocene sea-level records. Nat Geosci 4:328–332

Renne PR et al (2013) Time scales of critical events around. The cretaceous-Paleogene boundary. Science 339:684–687

Rhiannon S (2016) Antarctic research into warm water causing rapid glacial ice melt. ABC News. http://www.abc.net.au/news/2016-12-17/search-for-source-of-warm-water-melting-totten-glacier/8129202

Rignot E et al (2011) Acceleration of the contribution of the Greenland and Antarctic ice sheets to sea level rise. Geophys Res Lett 38(5):L05503

Roberts EM et al (2016) Oligocene termite nests with in situ fungus gardens from the Rukwa Rift Basin, Tanzania, support a Paleogene African origin for insect agriculture. Plos One. http://journals.plos.org/plosone/article?id=10.1371/journal.pone.0156847

Roe G (2006) In defense of Milankovitch. Geophys Res Lett 33:L24703

Rohling EJ et al (2014) Sea-level and deep-sea-temperature variability over the past 5.3 million years. Nature 508:477–482

Royer DL (2006) CO2-forced climate thresholds during the Phanerozoic. Geochim Cosmochim Acta 70:5665–5675

Royer DL, Berner RA, Beerling DJ (2001) Phanerozoic atmospheric CO change: evaluating geochemical and paleobiological approaches. Earth-Sci Rev 54:349–392

Royer DL, Berner RA, Montañez I, Neil P, Tabor J, Beerling DJ (2004) CO 2 as a primary driver of Phanerozoic climate. GSA Today 14:3

Royer DL, Berner RA, Park J (2007) Climate sensitivity constrained by CO 2 concentrations over the past 420 million years. Nature 446:530–532

Rubens CL, Figueira EI, Cunha L (1998) Contamination of the oceans by anthropogenic radionuclides. Quím Nova 21:73–77

Ruddiman WF (1997) Tectonic uplift and climate change. Springer, New York, p 380

Ruddiman WF (2003) The anthropogenic greenhouse Era began thousands of years ago. Climate Change 61:261–293

Ruff TA (2015) The humanitarian impact and implications of nuclear test explosions in the Pacific region. Int Rev Red Cross 97(899):775–813

Ryall J (2011) New radiation hotspots prompt Japan to extend monitoring. The Telegraph. http://www.telegraph.co.uk/news/worldnews/asia/japan/8721630/New-radiation-hotspots-prompt-Japan-to-extend-monitoring.html

Sagan C (1980) Cosmos. Random House, New York, p 396

Sagan C (1983) The Nuclear winter: the World after Nuclear War. http://www.e-reading.club/bookreader.php/148584/Sagan_-_The_Nuclear_Winter___The_World_After_Nuclear_War.pdf

Schell J (2000) The fate of the Earth/the abolition. Stanford University Press, Redwood City, p 460

Schlosser E (2016) World war three by mistake: harsh political rhetoric, combined with the vulnerability of the nuclear command-and-control system, has made the risk of global catastrophe greater than ever. The New Yorker December 23, 2016. http://www.newyorker.com/news/news-desk/world-war-three-by-mistake

Schoene B (2009) Correlating the end-Triassic mass extinction and flood basalt volcanism at the 100 ka level. Geology 38:387–390

Schopf JW (2001) Cradle of life: the discovery of Earth's earliest fossils. Princeton University Press, Princeton, p 361

Scott MR, Salter PF, Halverson JE (1983) Transport and deposition of plutonium in the ocean: evidence from Gulf of Mexico sediments. Earth Planet Sci Lett 63:202–222

Sepkoski JJ (1998) Rates of speciation in the fossil record. Phil Trans R Soc Lond B353:315–326

SETI Institute (2016) The Drake Equation: what do we need to know about to discover life in space? http://www.seti.org/drakeequation

Shakun JD, Clark PU, He F, Marcott SA, Mix AC, Liu Z, Otto-Bliesner B, Schmittner A, Bard E (2012) Global warming preceded by increasing carbon dioxide concentrations during the last deglaciation. Nature 484:49–55

Sherwood V (1996) Book of insects records. University of Florida, Gainesville. Chapter 1: Most heat tolerant

Siegenthaler U et al (2005) Stable carbon cycle–climate relationship during the late Pleistocene. Science 310:1313–1317

Simon SL et al (2010) Radiation doses and cancer risks in the Marshall islands associated with exposure to radioactive fallout from Bikini and Enewetak nuclear weapons tests: summary. Health Phys 99(2):105–123

Simone D et al (1979) The effects of nuclear war. Nuclear War Effects Project Library of Congress Catalog Card Number 79–600080. http://atomicarchive.com/Docs/pdfs/7906.pdf

Sjoeblom K Linsley GS (1993) International Arctic Seas Assessment Project (IASAP). http://www.iaea.org/inis/collection/NCLCollectionStore/_Public/28/041/28041350.pdf

Smith SJ et al (2011) Anthropogenic sulfur dioxide emissions:1850–2005. Atmos Chem Phys 11(1116):2011

Smith JN et al (2015) Arrival of the Fukushima radioactivity plume in North American continental waters. Proc Natl Acad Sci 112(5):1310–1315

Sodhi NS, Brook BW, Bradshaw JA (2009) Causes and consequences of species extinctions. Princeton guide to Ecology, p 514. http://connection.ebscohost.com/c/book-chapters/53752656/chapter-v-1-causes-consequences-species-extinctions

Solanki SK (2002) Solar variability and climate change: is there a link? A&G 43(5):5.9–5.13

Solomon S, Plattner GK, Knutti R, Friedlingstein P (2009) Irreversible climate change due to carbon dioxide emissions. Proc Natl Acad Sci 106(6):1704–1709

Species Extinction – the facts. Redlist. http://cmsdata.iucn.org/downloads/species_extinction_05_2007.pdf

Stanley SM (2016) Estimates of the magnitudes of major marine mass extinctions in earth history. Proc Natl Acad Sci 113:E6325–E6334

Starr S (2015) Nuclear War, Nuclear Winter, and Human Extinction. https://fas.org/pir-pubs/nuclear-war-nuclear-winter-and-human-extinction/

Steffen W, Crutzen J, McNeill JR (2007) The Anthropocoene: are humans now overwhelming the great forces of nature? Ambio 36(8):614–621

Steffen W et al (2016) Stratigraphic and Earth System approaches to defining the Anthropocene. Earth Future 4:324–345

Steffensen JP et al (2008) High-resolution Greenland ice core data show abrupt climate change happens in few years. Science 321:680–684

Sumner T (2016) Bikini Atoll radiation levels remain alarmingly high. Sci News 190(1):16

Taylor A (2015) 70 years since trinity: the day the nuclear age began. July 16, 2015. https://www.theatlantic.com/photo/2015/07/70-years-since-trinity-when-we-tested-nuclear-bombs/398735/

TeleSUR United States Has Enough Operational Nuclear Weapons to Destroy Itself 4 Times. http://www.telesurtv.net/english/news/US-Has-Enough-Nuclear-Weapons-to-Destroy-Itself-4-Times-Over---20140914-0033.html

Termite Biology. https://www.termatrac.com/home-owners/learn-about-termites/termite-biology/

Termites: how do ants incredibly survive even after cooking in microwave oven? https://www.quora.com/How-do-ants-incredibly-survive-even-after-cooking-in-microwave-oven

The Chernobyl Forum (2003–2005) Chernobyl's legacy: health, environmental and socio-economic impacts and recommendations to the Governments of Belarus, the Russian Federation and Ukraine. https://www.iaea.org/sites/default/files/chernobyl.pdf

The Chernobyl Gallery: photographs of Chernobyl now, 30 years after the world's worst nuclear disaster. A virtual tour of Pripyat in pictures from the safety of your screen. http://chernobylgallery.com/

The Cradle of Civilization. https://www.khanacademy.org/humanities/ancient-art-civilizations/ancient-near-east1/the-ancient-near-east-an-introduction/a/the-cradle-of-civilization

The Extinction Crisis. Centre for Biological Diversity. http://www.biologicaldiversity.org/programs/biodiversity/elements_of_biodiversity/extinction_crisis/

The Guardian (2014) Marshall Islands Bikini Atoll nuclear test: 60 years later and islands still unliveable. https://www.theguardian.com/world/2014/mar/02/bikini-atoll-nuclear-test-60-years

The Manhattan Project: an interactive history. U.S. Department of Energy. https://www.osti.gov/opennet/manhattan-project-history/Events/1945/trinity.htm

The US Atlantic Meridional overturning circulation program. Climate variability and predictability program. https://usclivar.org/amoc

Tilyard RJ (1937) Kansas Permian insects. Part XX the cockroaches, or order Blattaria. Am J Sci 34:169–202. (249–276)

Turco RP et al (1983) Nuclear winter: global consequences of multiple nuclear explosions. Science 222:1283–1292

Turner JS (2011) Termites as models of swarm cognition. Swarm Intell 5:19–43

Turner JS (2015) The 'Collective Mind' of the Termite. Live Science October 30, 2015. http://www.livescience.com/52644-the-collective-mind-of-the-termite.html

Tyrrell KA (2017) Abrupt climate change could follow collapse of Earth's oceanic conveyor belt. University of Wisconsin–Madison News. http://news.wisc.edu/abrupt-climate-change-could-follow-collapse-of-earths-oceanic-conveyor-belt/

UK Meteorological Office (2011) 25 years on from Chernobyl. https://blog.metoffice.gov.uk/2011/04/26/25-years-on-from-chernobyl/

Union of Concerned Scientists – Climate Hot Map: Global Warming Effects Around the World. http://www.climatehotmap.org/global-warming-locations/western-siberia.html

Union of Concerned Scientists (2016) The risk of Nuclear War with China 2016. http://www.ucsusa.org/nuclear-weapons/us-china-relations/risk-nuclear-war-china#.WKzYKbm7oy9

United Nation Office of Disarmament Affairs. https://www.un.org/disarmament/

United States Has Enough Operational Nuclear Weapons to Destroy Itself 4 Times. http://www.telesurtv.net/english/news/US-Has-Enough-Nuclear-Weapons-to-Destroy-Itself-4-Times-Over---20140914-0033.html

US Geol Surv (2016) PRISM, Pliocene Research, interpretation and Synoptic Mapping. http://geology.er.usgs.gov/egpsc/prism/

Van der Sloot B (2016) Welcome to the Jungle: the liability of internet intermediaries for privacy violations in Europe. https://www.jipitec.eu/issues/jipitec-6-3-2015/4318

Vaughan A (2015) Humans creating sixth great extinction of animal species, say scientists. https://www.theguardian.com/environment/2015/jun/19/humans-creating-sixth-great-extinction-of-animal-species-say-scientists

Velicogna I (2009) Increasing rates of ice mass loss from the Greenland and Antarctic ice sheets revealed by GRACE. Geophys Res Lett 36(19):L19503

Velicogna I (2014) Regional acceleration in ice mass loss from Greenland and Antarctica using GRACE time-variable gravity data. Geophys Res Lett 41:8130–8137

Velicogna I, Wahr J (2005) Greenland mass balance from GRACE. Geophys Res Lett 32:L18505

Veron C (2008) Mass extinctions and ocean acidification: biological constraints on geological dilemmas. Coral Reefs 27:459–472

Veron JEN (2009) The coral reef crisis: the critical importance of <350 ppm CO2. Mar Pollut Bull 58:1428–1436

Vitousek PM (1994) Beyond global warming: Ecology and global change. Ecology 75(7):1861–1876

Vizcarra N (2014) Strange bedfellows: Is the delicate relationship between fire and plants changing? Earthdata. https://earthdata.nasa.gov/user-resources/sensing-our-planet/strange-bedfellows-2014

Wagner F, Aaby B, Visscher H (2002) Rapid atmospheric CO 2 changes associated with the 8,200-years-B.P. Cooling event. Proc Natl Acad Sci U S A 99:12011–12014

Ward PD (2007) Under a green sky: global warming, the mass extinctions of the past, and what they can tell us about our future. Hareer and Collins, New York, p 205

Ward P (2008) A theory of Earth'sMass extinctions (TED2008 19:41 Filmed Feb 2008). https://www.ted.com/talks/peter_ward_on_mass_extinctions

Ward P (2015) The Medea hypothesis: is life on earth ultimately self-destructive? Princeton University Press, Princeton, p 208
Wax H (2008) Using math to explain how life on earth began. Sci Am October 2008. https://www.scientificamerican.com/article/can-math-solve-origin-of-life/
Wells HG (2011) The War of the Worlds. Penquin, London, p 240. http://www.shmoop.com/war-of-the-worlds-hg-wells/summary.html
Word Building (2016) How many survivors would be left after a global nuclear war? http://worldbuilding.stackexchange.com/questions/28335/how-many-survivors-would-be-left-after-a-global-nuclear-war
World Nuclear Association (2012) Radiation and life. http://www.world-nuclear.org/information-
World Nuclear Association (2017). http://www.world-nuclear.org/
World Nuclear Weapon Stockpile. http://www.ploughshares.org/world-nuclear-stockpile-report
Wrangham R (2009) Catching fire: how cooking made us human. Basic Books, New York, p 320
Wright JD, Scahller MF (2013) Evidence for a rapid release of carbon at the Paleocene-Eocene thermal maximum. Proc Natl Acad Sci 110(40):15908–15913
Wu J et al (2014) Isotopic composition and Distribution of plutonium in Northern South China Sea sediments revealed continuous release and transport of Pu from the Marshall Islands. Environ Sci Technol 48(6):3136–3144
WWF (2016a) – How many species are we losing? http://wwf.panda.org/about_our_earth/biodiversity/biodiversity/
WWF (2016b) – Impact of habitat loss on species. http://wwf.panda.org/about_our_earth/species/problems/habitat_loss_degradation/
Wynn G (2014) How much worse is a 4 degrees world? Climate Home. http://www.climatechangenews.com/2014/04/01/how-much-worse-is-a-4-degrees-world/
Xu L et al (2013) Temperature and vegetation seasonality diminishment over northern lands. Nat Clim Chang 3:581–586
Yablokov AV, Nesterenko VB (2009) Chernobyl contamination through time and space. https://www.ncbi.nlm.nih.gov/pubmed/20002040
Yokoyama Y, Esat TM (2011) Global climate and sea level: enduring variability and rapid fluctuations over the past 150,000 years. Oceanography 24:54–69
Zachos J, Pagani M, Sloan L, Thomas E, Billups K (2001) Trends, rhythms, and aberrations in global climate 65 Ma to present. Science 292:686–693
Zachos J, Dickens GR, Zeebe RE (2008) An early Cenozoic perspective on greenhouse warming and carbon-cycle dynamics. Nature 451:279–283
Zeebe RE, Bada JL, Zachos JC, Dickens GR (2009) Carbon dioxide forcing alone insufficient to explain Paleocene–Eocene thermal maximum warming. Nat Geosci 2:576–580
Zeebe RE, Ridgwell A, Zachos JC (2016) Anthropogenic carbon release rate unprecedented during the past 66 million years. Nat Geosci 9:325–329
Zheng J et al (2012) Isotopic evidence of plutonium release into the environment from the Fukushima DNPP accident. Scientific Reports 2, Article number: 304. http://www.nature.com/articles/srep00304
Zheng J et al (2013) Release of plutonium isotopes into the environment from the Fukushima Daiichi nuclear power plant accident: what is known and what needs to be known. Environ Sci Technol 47(17):9584–9595

Index

A.Y. Glikson, *The Plutocene: Blueprints for a Post-Anthropocene Greenhouse Earth*, Modern Approaches in Solid Earth Sciences 13,
DOI 10.1007/978-3-319-57237-6

MIX
Papier aus verantwortungsvollen Quellen
Paper from responsible sources
FSC® C105338

If you have any concerns about our products,
you can contact us on
ProductSafety@springernature.com

In case Publisher is established outside the EU,
the EU authorized representative is:
Springer Nature Customer Service Center GmbH
Europaplatz 3, 69115 Heidelberg, Germany

Printed by Libri Plureos GmbH
in Hamburg, Germany